Solar Hydrogen
Fuel of the Future

Solar Hydrogen
Fuel of the Future

Mario Pagliaro
CNR, Institute of Nanostructured Materials and Institute for Scientific Methodology, Palermo, Italy

Athanasios G. Konstandopoulos
CPERI/CERTH and Aristotle University, Thessaloniki, Greece

RSCPublishing

ISBN: 978-1-84973-195-9

A catalogue record for this book is available from the British Library

Published by The Royal Society of Chemistry,
Thomas Graham House, Science Park, Milton Road,
Cambridge CB4 0WF, UK

Registered Charity Number 207890

For further information see our web site at www.rsc.org

Printed in the United Kingdom by Henry Ling Limited, at the Dorset Press, Dorchester, DT1 1HD

Dedication

This book is dedicated to Rosaria and Eta for being there all this time

Preface

Stay away from books that appear in fashion-chasing clusters: excellent recent examples are books on nano-everything, on hydrogen economy and on global warming.[1]

-V. Smil

This warning advice from Vaclav Smil, a leading authority in energy science, has been a constant guiding admonition on writing this book.

Using plain language, numerous illustrations and updated practical examples, in the following chapters we aim to show that hydrogen produced using solar energy to split water *is* the energy currency of the future because it solves two key requirements of our global society – namely the incessant flow of energy and clean energy – and does so at an affordable cost.

In other words, when produced from renewable energy and renewable raw materials – namely sunlight and water – hydrogen is an entirely clean and powerful fuel capable of replacing climate-altering fossil fuels to power homes, trucks, boats and turbines, including thermoelectric power units. Indeed, in reacting with air's oxygen, hydrogen releases a huge amount of energy and does so without any harmful emissions into the environment, because pure water is the only by-product of the reaction.

What is now required to make this fuel widely and readily available is to find ways to make it locally available to users, without having to wait for the building of hydrogen pipelines and hydrogen fuel station infrastructures.

Solar Hydrogen: Fuel of the Future
Mario Pagliaro and Athanasios G. Konstandopoulos
© Mario Pagliaro and Athanasios G. Konstandopoulos 2012
Published by the Royal Society of Chemistry, www.rsc.org

The first good news in this book is that these technologies will eventually be available on the market at an affordable cost.

The second is that accelerated innovation in both major domains of solar energy technology (photovoltaics and concentrated solar power) has resulted in the rapid fall of the solar electricity price, opening the route to a number of practical applications using solar hydrogen energy.

In detail, direct thermochemical water splitting using concentrated solar power (CSP, Chapter 3) has the potential to store solar energy and convert it into clean hydrogen, using a tiny fraction of the world's area to meet our present and future global energy needs.

Photovoltaic (PV, Chapter 2) energy, in its turn, has the versatility required to support the creation of a distributed energy generation infrastructure based on hydrogen energy, in both economically developed *and* developing countries, especially now that the price of PV solar electricity has fallen to historically low levels.

The sun delivers 5000 times our present global power needs. Furthermore, in contrast to what some people might tend to think, solar energy is evenly distributed. Thus, in contrast to the reservoirs of crude oil or natural gas, solar hydrogen is *not* amenable to exploitation by oligopolies. However, although it is inexhaustible and evenly distributed, sunlight, like wind, is intermittent. Hence, for widespread use of solar power we need to store it efficiently.

We therefore need to use solar hydrogen, because most storage media employed thus far – batteries, ultracapacitors and pumped-storage hydroelectric – all have significant disadvantages, such as self-discharge, high cost, and the use of toxic and rare materials in their construction.

In a recent paper,[2] focusing on economic convenience only, Derek Abbott has suggested that sunlight is the *scalable* source of power on which our future energy needs must rely. Notably, low-tech CSP (concentrated solar thermal power) is preferred to photovoltaic solar cells. Using mirrors (heliostats) to focus sunlight on a solar thermal collector, an area as little as a 500 by 500 km^2 (a tiny fraction of the world's desert area) will suffice to supply the world's energy needs.

For Abbott, simple combustion of hydrogen should be preferred to both electric batteries and hydrogen fuel cells, because the latter two would be not scalable owing to the use of expensive membrane technology, as well as expensive metal catalysts (platinum) or conducting metal ions (lithium), respectively.

However, we argue, new generation fuel cells are nanochemistry-based (Chapter 4), and they either require far lower amounts of noble metals, or will simply use low cost metal catalysts such as nickel, to generate power at levels of efficiency that are simply out of scope for Carnot-limited systems such as the internal combustion engine.

In brief, hydrogen energy finally makes sense, even if there are efficiency losses in compressing and delivering gaseous H_2, because the available solar power is virtually unlimited and the costs of all the technologies required – CSP, PV and water electrolysis (Chapter 2) – are becoming low.

For example, the intrinsic versatility and the current low cost of PV electricity supports the option of using solar PV stations to produce electricity locally, as well as relatively small amounts of H_2, to refuel and power cars or boats powered by fuel cells.

If a $250\,m^2$ PV-electrolysis solar station, for example the one shown in Chapter 2 in Austria, can produce $823\,kg$ of pure H_2 per year, in regions like Sicily where PV electricity has already reached grid-parity,[3] this figure should be doubled at a fraction of the cost.

By the same logic, as millions of people in countries of Africa and Asia start using electricity, they will soon use solar hydrogen in place of lead batteries to build a truly widespread power distribution infrastructure, in which each home can produce and store enough electricity from the sun to be self-sufficient.

Not much appears to have changed since 2007 when the International Association for Hydrogen Energy submitted a memorandum to the heads of State at the G8 summit of 2007 (and, again, in 2009 in Italy), asking them to give hydrogen energy top priority. The memorandum read:

> Hydrogen energy: the abundant clean energy for humankind as a means of mitigating anthropogenic climate change, avoiding environmental challenges, and decelerating the world's ongoing oligopolization of conventional energy raw materials is the permanent solution to the upcoming energy and climate change catastrophe.[4]

Attentive readers, however, should not be misled by this apparent lack of significant achievements in the last five years. As the same readers will notice, a number of new hydrogen-powered devices have quietly entered the market, along with solar hydrogen generators. The former products indeed run on water and electricity only, to generate and safely store solar hydrogen that is later used as fuel.

Users can now opt for a solar hydrogen cartridge (Chapter 4), for instance, to power their portable electronic devices. Just add water to the energy filling station and no more rechargeable batteries will be needed. Plagued by inefficient battery technology, several manufacturers of portable electronic devices are considering eventually switching to hydrogen fuel cells to power their phones.[5]

Similarly, as readers will realize on reading Chapter 4, there already are homes whose owners simply use an electrolyzer and PV modules to self-generate all the fuel required to power their household appliances. In Italy, the first urban hydrogen pipeline is operating safely and has been supplying customers with clean energy for three years, while a 12 MW power plant burns hydrogen to generate electricity and heat.

These changes are apparently minor, in view of the several terawatt scale that is needed to make a substantial impact on the global fuel market. However, people will soon become accustomed to hydrogen energy and this will convince them that hydrogen can be safely handled with huge technical, economic and environmental benefits.

After this, widespread change will be initiated. People will start to look for cars, boats and mobile vehicles in general that will run on solar hydrogen. Those companies wise enough to have invested in this new power technology will be the first to harvest the benefits of the massive demand for non-fossil fuels that is already present in the market, even in affluent countries.

Indeed, despite the global economic recession that started in 2008, the price of oil consistently remains >90 US$/barrel. This can only accelerate the transition to solar energy fuel because many countries need to reduce imports of foreign oil and natural gas, which even in a relatively small country like Italy cost some 60 billion Euros per year.

Hydrogen energy, clearly, is a hot research and industrial topic. In just the last two years there have been substantial breakthroughs in H_2 generation, as well as in water oxidation catalysts, from the laboratories of Long,[6] Chorkendorff,[7] Hill[8] and Nocera.[9]

Websites such as *hydrogenfuelnews.com* report daily on hydrogen powered cars and the use of commercial hydrogen fuel cells to power homes, industries and transportation. Similarly, excellent volumes have been published recently that survey solar hydrogen energy production via traditional,[10] photoelectrolytic[11] or nanotechnology based[12] methods.

Hydrogen storage[13] can occur in different classes of materials, including metal hydrides, inorganic porous solids, organic materials, nanotubes and, most recently, graphene. Yet, as Frauscher's *Riviera 600* hydrogen boat or the Honda *FCX Clarity* automobile clearly show, hydrogen can now be stored at sufficiently high density for use in commercial mobile vehicles.

Ongoing research in solar thermochemistry has also provided promising results to support the use of a combined hydrogen and carbon storage solution, namely that of carbon neutral solar hydrocarbons, as an energy carrier compatible with present day technology and that will provide an evolutionary bridge to a future hydrogen energy economy.

This book is unique in that it provides a critical and hopefully balanced insight into solar hydrogen energy that focuses on two main technologies, direct water splitting and PV-based electrolysis, which were selected on the basis of their practical relevance. Rather than surveying existing and emerging methodology, the goal of this book is to describe the near term implementation of H_2 technology, based on existing materials components and conceptual system designs.

Such a discussion of critical issues, which decision makers and policy makers (both in industry and in government) need to consider, is, in our opinion, timely and important. Solar hydrogen is *the* crucial technology for our common future. It is often claimed, for instance, that components of existing electrolyzers, such as catalysts for H_2 generation (noble metals like Pt), are not sufficiently abundant on Earth to provide the amounts needed to build the massive number of electrolyzers required for the large scale generation of solar H_2, which is on the order of several terawatts.

If this were the case, the technology to generate solar hydrogen at the terawatt level would not yet exist. However, as detailed in Chapter 2, emerging commercial electrolyzers make use of cheap and readily available nickel catalysts, entirely obviating the use of any platinum or palladium catalysts. Thus, it may be concluded that there *does* already exist a technology suitable for terawatt level generation of solar hydrogen.

We hope that the readership of this book will not be limited to practitioners in the field (faculty members and students in chemistry, engineering and materials science) but will include decision makers, entrepreneurs, managers and professionals who need to increase their knowledge of solar hydrogen energy technology, and especially how it will open new opportunities for growth in all countries and how it will influence various industries.

It is of course our hope that the book will also be of interest to engineers and researchers who are not experienced in solar hydrogen but would like to learn more about its prospects for the future.

Mario Pagliaro, Athanasios G. Konstandopoulos
Palermo and Thessaloniki

References

1. G. Ross, Scientists' Nightstand: Vaclav Smil, *American Scientist*, www.americanscientist.org/bookshelf/pub/vaclav-smil
2. D. Abbott, *Proc. IEEE*, 2010, **98**, 1931.
3. W. Hoffmann, President of the European Photovoltaic Industry Association, commenting on the study *SET for 2020* (2009). See also: http://greeninc.blogs.nytimes.com/2009/06/22/industry-group-says-solar-to-become-cost-competitive-in-italy-next-year/
4. International Association for Hydrogen Energy, *On Hydrogen Energy – The Forever Fuel. A Centennial Memorandum*, 2006. www.hydrogen.ru/modules/ContentExpress/img_repository/iaheCentennialMemo.pdf
5. Apple plots smartphones powered by hydrogen, *The Telegraph*, 25 December 2011.
6. H. I. Karunadasa, C. J. Chang and J. R. Long, *Nature*, 2010, **464**, 1329.
7. Y. Hou, B. L. Abrams, P. C. K. Vesborg, M. E. Björketun, K. Herbst, L. Bech, A. M. Setti, C. D. Damsgaard, T. Pedersen, O. Hansen, J. Rossmeisl, S. Dahl, J. K. Nørskov and I. Chorkendorff, *Nature Mater.*, 2011, **10**, 434.
8. Q. Yin, J. Miles Tan, C. Besson, Y. V. Geletii, D. G. Musaev, A. E. Kuznetsov, Z. Luo, K. I. Hardcastle and C. L. Hill, *Science*, 2010, **328**, 342.
9. M. W. Kanan and D. G. Nocera, *Science*, 2008, **321**, 1072.
10. K. Rajeshwar, R. McConnell and S. Licht (ed.), *Solar Hydrogen Generation*, Springer, Berlin, 2008.
11. C. A. Grimes, O. K. Varghese and S. Ranjan, *Light, Water, Hydrogen: The Solar Generation of Hydrogen by Water Photoelectrolysis*, Springer, Berlin, 2008.
12. L. Vayssieres (ed.), *On Solar Hydrogen and Nanotechnology*, John Wiley & Sons, New York, 2010.
13. A. Züttel, A. Borgschulte and L. Schlapbach (ed.), *Hydrogen as a Future Energy Carrier*, Wiley-VCH, Weinheim, 2008.

Acknowledgements

We would like to thank our collaborators, especially Giovanni Palmisano, Rosaria Ciriminna, Vittorio Loddo, Leonardo Palmisano and Vincenzo Augugliaro, for their generous collaboration in the activities of Sicily's PV Research Pole.

In addition, thanks are due to the European Commission for its support of the HYDROSOL research projects during the last decade, and to all past and present members of the HYDROSOL research consortiums, for their continued effort towards a clean energy future. In particular Chrysa Pagkoura and Souzana Lorentzou (both from the APT Lab at CPERI/CERTH) are gratefully acknowledged for their assistance with the present book.

Thanks are also due to Merlin Fox, Sue Humphreys, Juliet Binns and Saphsa Codling of RSC Publishing for their assistance in producing this book.

Mario Pagliaro, Athanasios G. Konstandopoulos

Contents

Solar Hydrogen: Fuel of the Future
Mario Pagliaro and Athanasios G. Konstandopoulos
© Mario Pagliaro and Athanasios G. Konstandopoulos 2012
Published by the Royal Society of Chemistry, www.rsc.org

About the Authors

Mario Pagliaro is a chemistry and management scholar based in Palermo at Italy's CNR where, since 2008, he has also led Sicily's Photovoltaics Research Pole. Mario's research focuses on the development of functional materials for a variety of uses and operates at the boundaries of chemistry and materials science. Between 1998 and 2003 he led the management educational center "Quality College del CNR", using the resulting income to establish a research group that currently collaborates with researchers in 11 countries. He has co-authored more than 100 scientific and technical papers as well as 12 scientific and management books, including two highly cited volumes, on silica-based materials and glycerol chemistry. In 2005 he was appointed *Maître de conférences associé* at the Montpellier Ecole Nationale Supérieure de Chimie and is frequently an invited speaker at scientific and technical meetings. Mario regularly organizes conferences and gives courses and tutorials on the topics of his research. In 2009 he chaired the 10th FIGIPAS Meeting in Inorganic Chemistry, held in Palermo, and in 2011 he was co-chairman of the first SuNEC international conference on solar energy. Currently he chairs the organizing committees of the SuNEC 2012 conference and of the FineCat 2012 symposium on catalysis for fine chemicals. Since 2004 he has organized the prestigious Seminar "Marcello Carapezza". His website is *qualitas1998.net*.

Athanasios G. Konstandopoulos is the Founder and has been the Director of the Aerosol & Particle Technology (APT) Laboratory at CPERI/CERTH (Thermi, Greece) since 1996. In 2006 he was elected as Director of CPERI and member of the Board of Directors of CERTH, and in 2011 he was elected Chairman of the Board and Managing Director of CERTH. He has also been a member of the faculty of Chemical Engineering at the Aristotle University since 2006, currently serving as Professor of New, Advanced & Clean Combustion Technologies. Dr Konstandopoulos has a hybrid background in Mechanical (Dipl. ME, Aristotle University of Thessaloniki, 1985; MSc ME Michigan Tech, 1987) and Chemical Engineering (MSc, MPhil, PhD, Yale University, 1991). He is a specialist in nanoparticles and combustion aerosols, with extensive research and engineering consulting experience in the design, modelling and testing of monolithic reactors for many applications including solar fuels (hydrogen and hydrocarbons) production, emission control for mobile and stationary applications and biotechnological applications. Dr Konstandopoulos is the leader of an international research team that has developed the HYDROSOL process for renewable hydrogen production by solar thermochemical water splitting. This achievement has received worldwide recognition through the 2006 Descartes Prize for Research (the highest scientific award in the European Union), the 2006 International Partnership for the Hydrogen Economy (IPHE) Inaugural Technical Achievement Award and the Global 100 Eco-tech Award at the 2005 Expo in Aichi, Japan. His extension of that work to include thermochemical splitting of carbon dioxide and its use for the synthesis of solar fuels, as an alternative to carbon storage technology, has won the highly competitive 2010 European Research Council (ERC) Advanced Grant. He has coordinated and managed more than 50 research projects, funded by the European Commission (EC) as well as leading international industries. He is the author of more than 160 scientific and technical papers, his research is widely cited and used by many academic and industrial research groups, and he is frequently an invited speaker at industry and scientific conferences. His research has been frequently covered by international communication media, including *Euronews Channel*, the *Financial Times*, *Die Welt*, *New Scientist* and *Chemistry World*, as well as by the domestic press. In 2005 he was elected one of the youngest ever Fellows of the Society of Automotive Engineers (SAE). In addition he has received the American Institute of Chemical Engineers First Place Award (1991) and the Yale University H. P. Becton

Prize for excellence in research (1992). Dr Konstandopoulos is currently a member of the Governing Board of the European Commission's Joint Research Centre (JRC), a member of several national and European committees in the area of energy research, and he serves as an evaluator for international research funding organizations. He has served as adjunct/ visiting professor at Universities in Europe and the USA, and he is, among others, a member of SAE, the Combustion Institute and the Gessellschaft fur Aerosolforchung, and a founding member of the Hellenic Association for Aerosol Research (elected to be its first president for 2006–2012). Updated information on his work can be found on *http://apt.cperi.certh.gr*.

CHAPTER 1

Hydrogen and Solar Hydrogen

1.1 Hydrogen: Structure and Properties

First produced by Robert Boyle in 1671 by reacting mineral acids with iron, hydrogen was recognized as a discrete substance by Henry Cavendish in 1766.[1] Cavendish named the gas "flammable air" and further reported in 1781 that it produced water when burned. In 1783, Lavoisier reproduced Cavendish's findings, formulating the mass conservation law, and named the element "hydrogen" from the Greek for "water-creator".[2]

At standard temperature and pressure, hydrogen is a colorless, odorless and non-toxic diatomic gas with the molecular formula H_2. It is the lightest (average atomic weight 1.007825 u for ^1H) and most abundant chemical element (75% of the Universe's elemental mass). Being highly reactive, the H_2 molecule on Earth is available only in chemical compounds and its extraction from said substances requires consumption of energy.

Hydrogen gas, furthermore, has an enormous volume (Table 1.1): 1 kg of H_2 at ambient temperature and atmospheric pressure has a volume of 11 m^3, therefore hydrogen storage for practical energy or chemical applications basically implies a reduction in volume. This is generally accomplished by increased pressure (in gas cylinders with a maximum pressure up to 80 MPa) or by liquefaction (achieved by lowering the temperature).[4]

We recall here that the boiling point of a pure substance increases with the applied pressure, up to a certain point. Propane, with a boiling point of –42 °C, for example, can be stored as a liquid at 21 °C under a moderate pressure of 7.7 bar. The boiling point of hydrogen can only be

Solar Hydrogen: Fuel of the Future
Mario Pagliaro and Athanasios G. Konstandopoulos
© Mario Pagliaro and Athanasios G. Konstandopoulos 2012
Published by the Royal Society of Chemistry, www.rsc.org

Table 1.1   Selected physical properties of hydrogen. (Reproduced from Ref. 3, with kind permission.)

Density	0.08988 g/L (0 °C, 101.325 kPa)
Melting point	14.01 K, −259.14 °C
Boiling point	20.30 K, −252.77 °C
Thermal conductivity	0.1805 W m^{-1} K^{-1}
Molar heat capacity	28.836 J mol^{-1} K^{-1}
Molecular weight	2.016 g mol^{-1}
Explosion limits in air	4.0–77.0 vol.%
Solubility in water	0.019 vol.% (0 °C, 101.325 kPa)
Minimum ignition energy	0.019 mJ
Specific heat capacity	14.199 kJ/kg K
Flame temperature in air	2318 K

increased to a maximum of −240 °C, through the application of approximately 13 bar, beyond which additional pressure has no beneficial effect. At standard conditions, hydrogen has a density of about 0.09 g L^{-1} while in liquid state its density increases to 70.8 g L^{-1} and its boiling point is only 20.3 K (−252.77 °C). Even as a liquid, therefore, hydrogen is not very dense. For comparison, every liter of water contains 111 kg of hydrogen, whereas a liter (cubic meter) of liquid hydrogen contains only 70.8 kg of hydrogen. Thus, water packs more mass of hydrogen per unit volume, because of its tight molecular structure, than hydrogen itself.

The phase diagram in Figure 1.1 indicates that liquid hydrogen exists only in a small region between the solid line and the line from the triple point at 21.2 K and the critical point at 32 K. This implies that once hydrogen is evaporated from liquid it is not possible to re-liquefy it by applying higher pressure, a method that works for many other gases. As shown in Table 1.1, hydrogen burns in air at a concentration in the range 4–77% by volume. The highest burning temperature of hydrogen, 2318 K, is reached at 29% concentration by volume. As little as 0.02 mJ is the minimum energy (thermal activation energy) required to ignite a stoichiometric hydrogen : oxygen mixture, which is one-tenth of the energy required to ignite a methane : oxygen mixture, for which the value is 0.29 mJ.

On a mass basis, the amount of energy produced during hydrogen combustion is higher than that released by any other fuel, with a low heating value (LHV, also known as net calorific value)[6] 2.4, 2.8 and 4 times higher than that of methane, gasoline and coal, respectively. Hydrogen indeed reacts easily with oxygen in a highly exothermic reaction (Equation 1.1) whose huge enthalpy is −286 kJ mol^{-1}.

$$2H_2(g) + O_2(g) \rightarrow 2H_2O(l) + 572 \, kJ \quad (286 \, kJ \, mol^{-1}) \qquad (1.1)$$

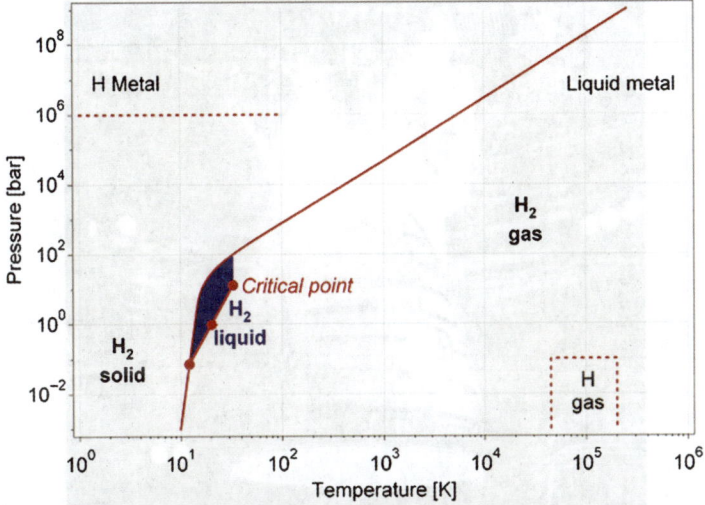

Figure 1.1 Primitive phase diagram for hydrogen. In hydrogen, the interaction between molecules is weak when compared with other gases, therefore the critical temperature is low ($T_c = 33.0$ K).
(Reproduced from Ref. 5, with kind permission.)

Table 1.2 Heating values for hydrogen and common hydrocarbon fuels. (Reproduced from Ref. 7, with kind permission.)

Fuel	Higher heating value (at 25 °C and 1 atm)	Lower heating value (at 25 °C and 1 atm)
Hydrogen	141.86 kJ g^{-1}	119.93 kJ g^{-1}
Methane	55.53 kJ g^{-1}	50.02 kJ g^{-1}
Propane	50.36 kJ g^{-1}	45.6 kJ g^{-1}
Gasoline	47.5 kJ g^{-1}	44.5 kJ g^{-1}
Diesel	44.8 kJ g^{-1}	42.5 kJ g^{-1}
Methanol	19.96 kJ g^{-1}	18.05 kJ g^{-1}

Table 1.2 shows that hydrogen has the highest energy-to-weight ratio of any fuel because hydrogen is the lightest element and has no heavy carbon atoms.

Therefore, for a given load duty, the mass of hydrogen required is only about one-third of the mass of hydrocarbon fuel needed. It is for this reason that hydrogen has been used extensively in the space program, where weight is crucial. For decades, for example, NASA has used liquid hydrogen to power space vehicles such as the Space Shuttle (Figure 1.2), chilling H$_2$ to near absolute zero (−252.87 °C), when hydrogen gas turns into a high-energy liquid. Compared with

Figure 1.2   The Space Shuttle main engine burns hydrogen with oxygen, producing a
nearly invisible flame at full thrust.
(Reproduced from Wikipedia.org, with kind permission.)

Table 1.3   Properties of hydrogen compared with those of other fuels.
(Reproduced from Ref. 8, with kind permission.)

Fuel	Boiling point [K]	Liquid density [kg m^{-3}]	Gas density [kg m^{-3}]
Hydrogen	20.3	71	0.08
Gasoline	350–400	702	4.68
Methanol	337	797	–
Methane	112	425	0.66
Propane	231	507	507
Ammonia	240	771	0.69

hydrocarbons, hydrogen gas has a good energy density by weight
$(33.3\,kWh\,kg^{-1})$ but a poor energy density by volume $(2.5\,kWh\,L^{-1})$,
which is about 3.5 times lower than the energy density by volume of
gasoline (Table 1.3).[8] Higher gas pressure values improve the energy
density by volume, allowing the tank to be smaller, but not lighter (at
least currently).

Alternatively, to increase its volumetric energy density, liquid hydrogen may be used. However liquefaction imposes large energy expenditure and the storage tanks must be well insulated to prevent boil-off.

The temperature of spontaneous ignition of hydrogen in air is high, *ca.* 500 °C. However, molecular hydrogen is highly flammable and will burn in air at a very wide range of concentrations, between 4% and 74% by volume, when the gas forms explosive mixtures with air. The mixtures spontaneously explode when triggered by spark, heat or sunlight. Pure hydrogen–oxygen flames emit ultraviolet light and are nearly invisible to the naked eye, as illustrated by the faint plume of the Space Shuttle main engine (see Figure 1.2) compared with the highly visible plume of the rocket boosters of the same shuttle.

The liquefaction process, involving pressurizing and cooling steps, is energy intensive. However, Equation 1.1 shows that pure, liquid water is the unique exhaust by-product of the reaction, and this fact alone – the absence of CO_2 emissions – immediately shows the enormous, benign environmental potential of the combustion of hydrogen when compared with combustion reactions employing hydrocarbons or coal.

Hydrogen indeed is an excellent fuel that can replace hydrocarbons with numerous advantages. For example, in 2009 the Italian power company Enel started operating a 12 MW H_2-powered electricity plant, in the industrial zone of Porto Marghera in Venice, which is fueled uniquely by hydrogen by-products from local petrochemical industries (Figure 1.3). The turbines were specially designed to resist embrittlement caused by hydrogen, but in any case the main emission of hydrogen combustion in air is water (the formation of trace amounts of nitrogen oxides is minimized by choosing the best combustion conditions).

Moreover, hydrogen burns much more efficiently and rapidly than gasoline, and replacement of gasoline with hydrogen fuel in automobiles is simple and fast, requiring substantial changes to the electronic and ignition systems only (Figure 1.4). Hydrogen tanks for storage can be installed in virtually any available space within the vehicle, and gaseous H_2 is precisely metered into the air intake of the engine.

Protium (^1H) is the most common hydrogen isotope, with an abundance of more than 99.98%. First prepared in 1932 by Urey,[9] deuterium (^2H or D, with an atomic weight of 2.0140, containing one proton and one neutron in its nucleus) is the other stable hydrogen isotope.

Given its relatively simple atomic structure, consisting only of a proton and an electron (Figure 1.5), the hydrogen atom, together with the spectrum of light produced from it or absorbed by it, has played a central role in the development of the theory of atomic structure culminating, in 1926, in the Schrödinger equation for the hydrogen

Figure 1.3   The first in the world to operate on such a scale, the 12 MW combined cycle plant in Venice's industrial area of Porto Marghera is fueled by hydrogen by-products from local petrochemical industries.
(Reproduced from http://criticae.wordpress.com, with kind permission.)

Figure 1.4   An existing vehicle engine can burn hydrogen or gasoline provided that the new hydrogen fuel system is activated using the same automobile gasoline fuel injection system.
(Photo courtesy of United Nuclear Scientific Supplies, reproduced from www.switch2hydrogen.com/h2.htm, with kind permission.)

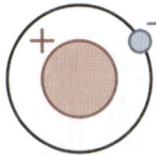

Figure 1.5 Protium, the most common isotope of hydrogen, has one proton and one electron. Unique among stable isotopes, it has no neutrons. (Reproduced from Wikipedia.org, with kind permission.)

atom.[10] Similarly, the corresponding simplicity of the hydrogen molecule was instrumental in allowing a fuller understanding of the nature of the chemical bond, which followed shortly after the quantum mechanical treatment of the hydrogen atom had been developed in the mid-1920s.

Differing by the relative spin of their nuclei, there are two different spin isomers of hydrogen diatomic molecules: *ortho-* and *para*hydrogen.[11] In the *ortho*hydrogen form, the spins of the two protons are parallel and form a triplet state with a molecular spin quantum number of 1 ($\frac{1}{2}+\frac{1}{2}$); in the *para*hydrogen form the spins are antiparallel and form a singlet with a molecular spin quantum number of 0 ($\frac{1}{2}-\frac{1}{2}$).

The *ortho* form is an excited state and has a higher energy than the *para* form. At standard temperature and pressure, the normal form of hydrogen gas contains about 25% of the *para* form and 75% of the *ortho* form. The *ortho*:*para* ratio in condensed H_2 is an important consideration in the preparation and storage of liquid hydrogen; the conversion from *ortho* to *para* is exothermic and produces enough heat to evaporate some of the hydrogen liquid, leading to loss of liquefied material. In general, rapidly condensed H_2 contains large quantities of the high-energy *ortho* form that converts to the *para* form very slowly.[12]

Hence, during hydrogen cooling, chemical catalysts (such as ferric oxide, activated carbon, platinized asbestos, rare earth metals, uranium compounds, chromic oxide, or some nickel compounds) are employed for the *ortho–para* interconversion.[13]

Molecular hydrogen is commonly used in power stations, as a coolant in generators, because its specific heat capacity is considerably higher than that of any other gas. Half a century before quantum mechanical theory was developed, Maxwell observed that the specific heat capacity of H_2 unaccountably departs from that of a diatomic gas below room temperature and begins increasingly to resemble that of a monatomic gas at cryogenic temperatures. This finding is ascribed by quantum theory to the wide spacing of the (quantized) rotational energy levels that results from the low mass of the molecule.[14] These widely spaced

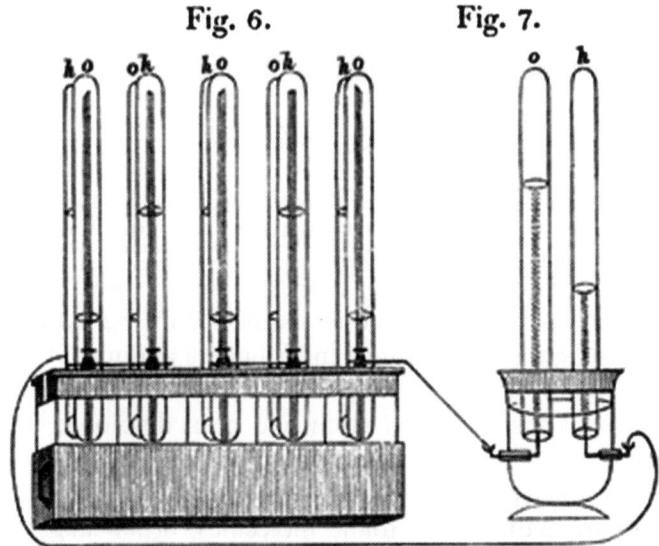

Figure 1.6   William Grove's drawing of an experimental "gas battery", from a letter
dated 1843.
(Reproduced from *Proceedings of the Royal Society*, with kind
permission.)

levels inhibit equal partition of heat energy into rotational motion in
hydrogen at low temperatures. Diatomic gases composed of heavier
atoms, such as N_2, do not have such widely spaced levels and do not
exhibit the same effect.[15]

Hydrogen can also react with oxygen to generate electricity in a "fuel
cell", a concept that was first proposed in 1838 by Schönbein. Soon
afterwards, in 1842, Grove introduced the first "gas voltaic battery",
using a design (Figure 1.6) that is remarkably similar to today's fuel cell
designs.[16] In contrast to batteries, which store and release chemical
energy as electricity until complete discharge, an H_2 fuel cell produces
energy by combining hydrogen with oxygen, continuing to function and
produce power as long as fuel and oxygen are supplied.

Liquid water is the only reaction product so that, for instance,
the toxic and polluting emissions produced by a boat's traditional
internal combustion engine, are replaced in a boat powered by a
hydrogen fuel cell by the emission of a benign flow of highly pure
water (Figure 1.7).

1.2   Hydrogen Production and Utilization

Hydrogen is widely employed in the chemical and petrochemical
industries. The total world production of hydrogen as a chemical

Figure 1.7   The only emission produced by Frauscher's Riviera 600 hydrogen-powered boat is clean water; H_2 is obtained cleanly by photovoltaic electrolysis of water.
(Photo courtesy of Fronius.)

constituent and as an energy source was valued at US\$120 billion in 2010 (Figure 1.8).[17] The two main applications are the production of ammonia (NH_3), via the Haber process, which is used directly or indirectly as fertilizer, and hydrocracking, namely converting heavy petroleum sources into lighter fractions suitable for use as fuels (and in the de-sulfurization of middle distillate diesel fuel). Furthermore, H_2 is used in the manufacture of methanol and hydrochloric acid, and as a hydrogenating agent, used particularly to increase the level of saturation of unsaturated fats and oils in the oleochemicals industry.

Finally, other important hydrogen users are the space flight business, the electronics industry, and metallurgical companies that require the reduction of metallic ores. In the latter cases, smaller quantities of "merchant" hydrogen are also manufactured and delivered to end users. The economies of scale inherent to large-scale oil refining and fertilizer manufacture make possible the on-site production and "captive" use of hydrogen.

Given that both the world's human population and the intensive agriculture used to support it are growing, the demand for ammonia is growing. The use of hydrocracking grows at an even faster rate, because

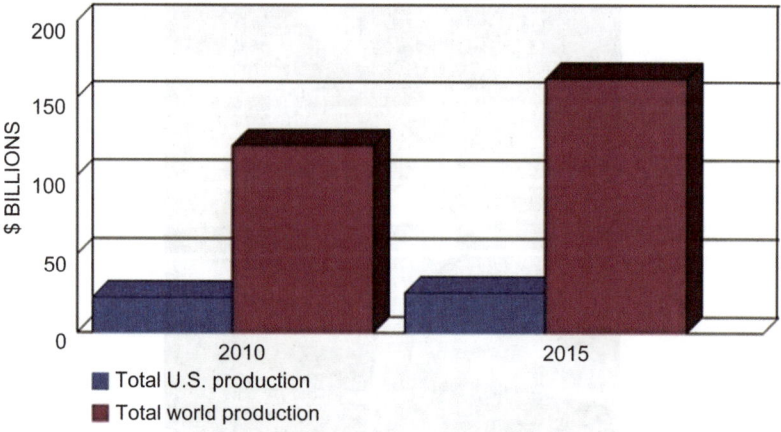

Figure 1.8   Hydrogen market and production in the world and in the USA, 2010 and
2015 (US$ billions).
(Reproduced from Ref. 17, with kind permission.)

Figure 1.9   Hydrogen is co-produced with carbon monoxide in natural gas "steam
reforming" plants such as this one.
(Photo courtesy of Linde.)

rising oil prices encourage oil companies to extract poorer source material, such as tar sands and oil shale. Accordingly, the hydrogen market
is expected to increase at a compound annual growth rate (CAGR) of
6.3%, to reach a value of US$163 billion in 2015 (see Figure 1.8).

Overall, around 96% of hydrogen is derived from fossil fuels and a
minor fraction is produced in large electrolyzers during the electrolysis
of brine. Most (49%) of the world's hydrogen is currently produced
from the steam reforming of natural gas (Figure 1.9), followed by

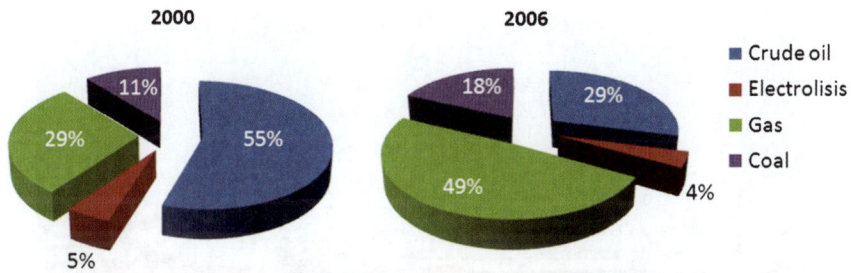

Figure 1.10 Worldwide hydrogen production in 2000 and 2006. (Reproduced from Ref. 12, with kind permission.)

Table 1.4 Hydrogen production in the chemical industry. (Adapted from Ref. 18, with kind permission.)

At high temperatures (700–1100 °C), steam (water vapor) reacts with methane to yield carbon monoxide and hydrogen. This reaction is favored at low pressures but is nonetheless conducted at 20 atm, because high-pressure hydrogen is the most marketable product. The "synthesis gas" product mixture is often used directly for the production of methanol and related compounds.

$$CH_4 + H_2O \rightarrow CO + 3H_2$$

Other important methods for H_2 production include partial oxidation of hydrocarbons:

$$2CH_4 + O_2 \rightarrow 2CO + 4H_2$$

and the coal reaction:

$$C + H_2O \rightarrow CO + H_2$$

Hydrogen is also manufactured via high or low pressure electrolysis of water, by decomposition of water (H_2O) into oxygen (O_2) and hydrogen gas (H_2) by means of an electric current passed through the water.

partial oxidation of oil (29%) and coal (18%). The graphs in Figure 1.10 show that the industrial manufacturing technology (Table 1.4) is rapidly changing, because in 2000 crude oil was the dominant source (55%).[18]

Given the huge and increasing volumes of H_2 produced in the world, it is perhaps no surprise that hydrogen plants are major energy-demanding processes and important releasers of, CO_2 emitting some 100 million tonnes of CO_2 equivalent per year.[19]

The current best electrolytic processes, on the other hand, have an efficiency of 50% to 80%, so that 1 kg of hydrogen (which has a specific

Figure 1.11   In the *Hydrogen Challenger* ship, based in Germany, hydrogen generated
by water electrolysis, induced by wind electricity generated on the open
sea, is stored and brought to shore where it can be injected into the
hydrogen infrastructure.
(Reproduced from seriouslygoodnews.com, with kind permission.)

energy of about 40 kWh kg^{-1}) requires 50 to 79 kWh of electricity. At
8 cents per kWh, this translates into \$4.00 per kg, which is, 3 to 10 times
the price of hydrogen from steam reformation of natural gas produced
using traditional methods.[20]

On the other hand, low cost renewable electricity is increasingly used
to manufacture H_2 electrolytically, for example in the German *Hydrogen
Challenger* ship (Figure 1.11). This vessel is equipped with vertical axis
wind turbines, which harness the strong winds of the outer seas and use
them to create the electricity used to electrolyze the water beneath the
ship.

Unlike electricity, hydrogen can be stored in large quantities and for
long periods of time. Storing energy in the form of hydrogen allows the
generator to sell power only when and where the price is highest. This
means that every kWh of hydrogen energy is much *more* valuable than a
kWh of electricity.

The use of hydrogen as a fuel is still a niche market. To date the
largest demand for hydrogen as a *fuel* has come from the United States
space program. Now, however, fuel cells that use hydrogen as a unique
fuel are opening up new markets for hydrogen suppliers, with potentially
high demand if some key applications take off (Figure 1.12).

Fuel cell vehicles (FCVs), for instance, involve technology under
development with five car makers (Daimler AG, Honda, General
Motors, Hyundai and Toyota) that are currently operating the largest
fleets of FCVs. According to a recent market report,[21] commercial sales

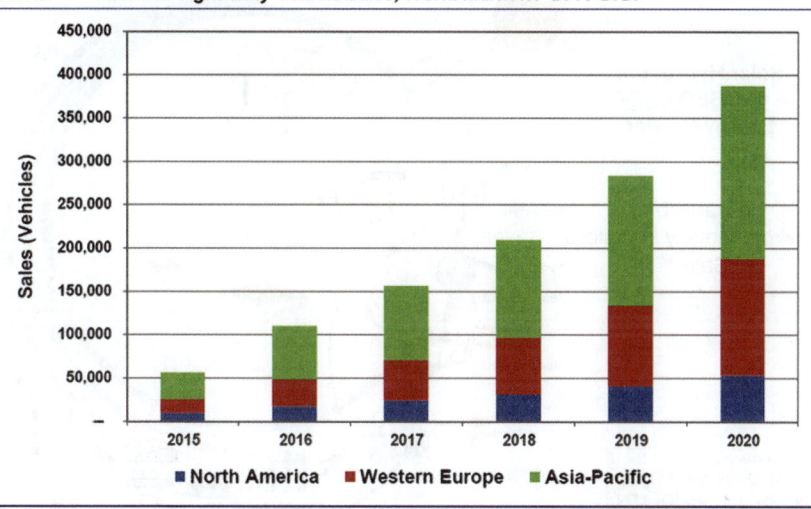

Chart 1.1   **Fuel Cell Light-Duty Vehicle Sales, World Markets: 2015-2020**

■ North America ■ Western Europe ■ Asia-Pacific

(Source: Pike Research)

Figure 1.12   The fuel cell car market is now in the ramp-up phase to commercialization, anticipated by automakers to occur around 2015. Following a pre-
commercialization period from 2010 to 2014, Pike Research forecasts
that 57 000 fuel cell vehicles (FCVs) will be sold in 2015, with sales
volumes increasing to 390 000 vehicles annually by 2020.
(Reproduced from Ref. 21, with kind permission.)

of FCVs will reach the key milestone of 1 million vehicles by 2020, with a
cumulative 1.2 million vehicles sold by the end of that year.

For the same US market analysts, the entire growth of the FCV
market is balancing on two key items: the growth of H_2 refueling
stations and the improved durability and efficiency of the fuel cells.
The growth of H_2 refueling stations will be boosted, we argue, by distributed solar hydrogen generation using photovoltaic electricity.

1.3   Solar Hydrogen

Hydrogen generated by water splitting induced by solar energy is the
fuel of the future, which can replace fossil fuels and ultimately cease our
dependence (or "addiction",[22] to quote a former US president) on fossil
hydrocarbons and coal, thus ending the emission of CO_2 into the
atmosphere that causes global warming and climate change.

In other words, clean solar-based H_2 technologies not only produce
hydrogen but also employ entirely renewable and abundant energy
sources and raw materials: solar energy and water, respectively, which
produce no CO_2 emissions (Figure 1.13).

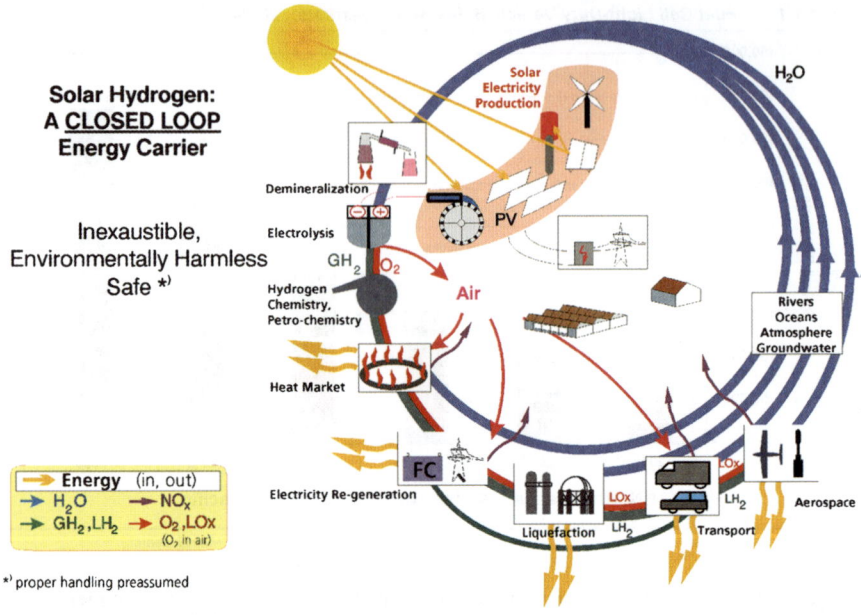

Figure 1.13   When made from solar energy and water, hydrogen closes the anthropogenic energy cycle, and turns into an entirely clean energy carrier.
(Reproduced from Ref. 23, with kind permission.)

In the words of Winter, a long-time advocate of hydrogen energy:[23]

> The solar water-to-hydrogen-to-water cycle is the only closed
> material cycle of any human energy scheme. All the others are open
> systems: they take something irrecoverable from the earth's crust,
> convert it chemically or nuclearly, and return it to the biosphere,
> sometimes toxic, sometimes radioactive, and sometimes of negative
> environmental or climatic influence.

In the last decade, much hype has been associated with the topic of
hydrogen. For example, in a widely read book,[24] Romm has questioned
the idea that hydrogen is an economically viable fuel for transportation
because of its cost and the greenhouse gases generated during production, the low energy content per volume and weight of the container, the
cost of the fuel cells, and the cost of the infrastructure.

In contrast to fossil fuel deposits, which are a concentrated source of
high-quality energy, commonly extracted with power densities (the rate
of energy production per unit of Earth's area) of 10^2 or 10^3 W m^{-2} for
coal or hydrocarbon fields, biomass energy production has densities well

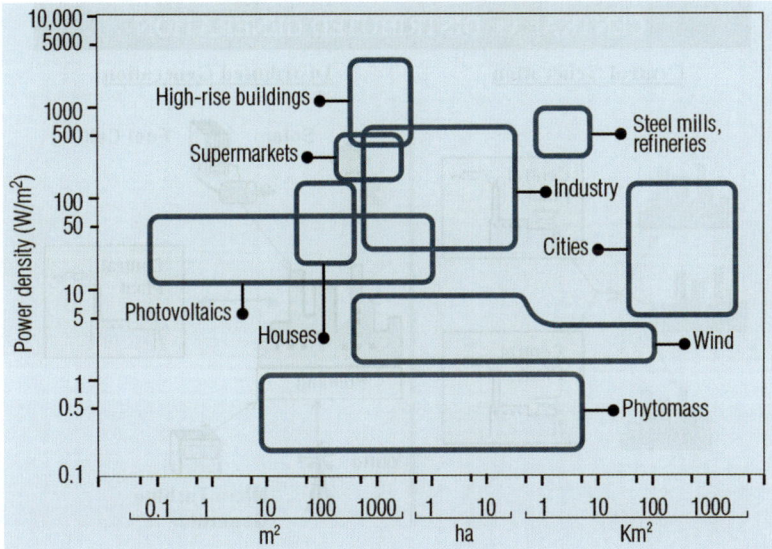

Figure 1.14 Power densities for renewable fuels and energy consumers. Power density is the rate of energy production per unit of the Earth's area, expressed in watts per square meter ($W\,m^{-2}$).
(Reproduced from Ref. 25, with kind permission.)

below $1\,W\,m^{-2}$, while the density of electricity produced by photovoltaic generation is around $20\,W\,m^{-2}$ of peak power.[25]

Figure 1.14 indeed shows that the energy supply chain of today's fossil-fueled civilization works by producing fuels and thermal electricity with power densities that are one to three orders of magnitude higher than the common power densities consumed by our buildings and cities. In other words, solar energy is the *only* renewable energy source with the versatility to meet both intensive production needs (through concentrated solar power, or CSP) and distributed energy demand through on-site installed photovoltaic (PV) modules, whose price in the last three years has fallen by $>90\%$ (from 6 to less than $0.7\,€\,W^{-1}$).

Distributed generation (Figure 1.15) refers to power generation at the point of consumption. Generating power on-site, rather than centrally, eliminates the cost, complexity, interdependencies, and inefficiencies associated with transmission and distribution. In the past three decades, computers and telephones have become decentralized and wireless. In a similar way, as emphasized by Bradford,[26] to distributed computing (*i.e.* the personal computer) and distributed telephony (*i.e.* the mobile phone), distributed generation using solar energy shifts control to the consumer.

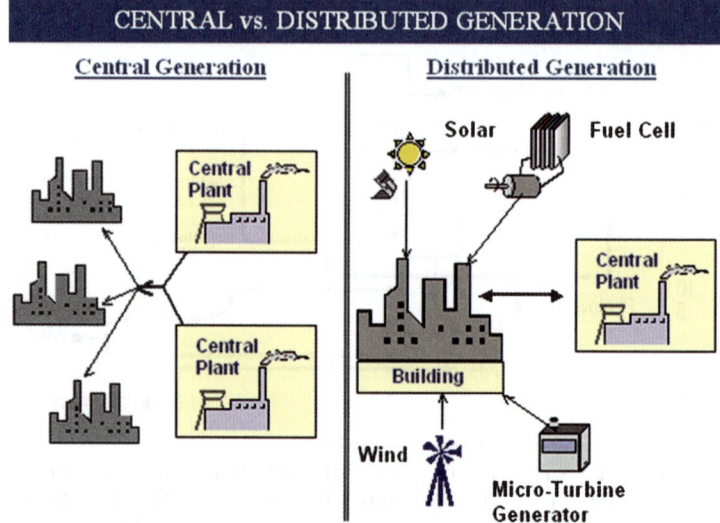

Figure 1.15   Distributed energy generation is based on local generation of energy
using microturbines, fuel cells, photovoltaic systems and wind power,
where energy is needed and consumed.

In other words, PV distributed on a small scale instead of on today's
industrial-size electricity grids will begin to compete with the economies
of scale that Edison's electricity transmission created over the last century. In the last five years the price of solar electricity has fallen to such
an extent that it has now reached the cost of fossil electric power in many
countries and regions, including southern Italy, Greece and Israel. This
means that distributed energy generation has become an economically
viable alternative to central energy generation in large fossil fuel power
stations.

Certainly, the shift from concentrated and easily transportable fuel
sources (oil converted to gasoline/diesel) to diffuse solar energy converted to less concentrated and less transportable hydrogen (Figure
1.16) fuel will require a substantial fall in the cost of H_2 fuel cells and H_2
fuel combustion engines.[27] Yet, as highlighted in this book, a number of
new technologies have already been marketed that make it possible to
generate and use solar hydrogen to power not only portable devices but
also cars, boats and homes.

In brief, solar H_2 is an excellent storage option for the *excess* solar
electricity generated during the day for use during night-time and cloudy
days. Two complementary methods exist, which rely on two major solar
energy technologies, namely photovoltaics (PV) and concentrated solar
power (CSP).

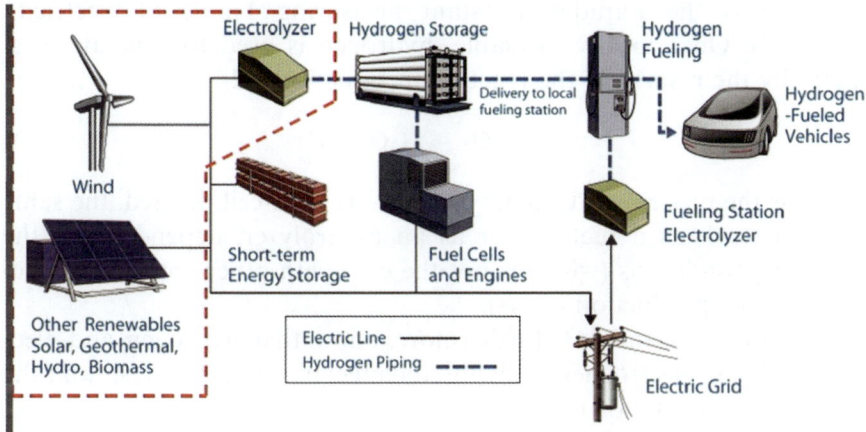

Figure 1.16 As the cost of solar electricity has reached grid-parity in many countries worldwide, the electrolysis of water to make solar hydrogen turns into an economically viable option for storing the free solar fuel radiation energy.
(Reproduced from www.theoildrum.com, with kind permission.)

Figure 1.17 Honda's solar-powered hydrogen production, storage and refuelling station in Torrance, California has been operating at the company's California laboratory since 2001.
(Reproduced from Ref. 29, with kind permission.)

The first technology involves water electrolysis using a photovoltaic current (Equation 1.2 and Figure 1.17).

$$H_2O + 2F \rightarrow H_2 + \tfrac{1}{2}O_2 \qquad (1.2)$$

where F is the Faraday constant measuring 1 mole of electricity (96 485 C). Once locally available, hydrogen is used to generate electricity by the reverse of the reaction in Equation 1.2:

$$H_2 + \tfrac{1}{2}O_2 \rightarrow H_2O + 2F \qquad\qquad (1.3)$$

which is the process that occurs in an H_2–O_2 fuel cell. Indeed, the same cell can work as a fuel cell or as an electrolyzer, depending on the operating conditions (which is not the case with other competing reactions for the production of H_2).

Equations 1.2 and 1.3, furthermore, show that hydrogen and electricity – being electrochemically interchangeable via electrolysis and the fuel cell – compete for the same primary solar energy.

In general, electrolytic hydrogen production is perfectly suited to distributed energy production, namely to create locally the amount of hydrogen required to power a house, a car or a boat's engine. This as occurs with most of the hydrogen refuelling stations in the USA,[28] such as the first one, which was opened in 2001 by Honda at its research and development center in Torrance, in the Los Angeles area (see Figure 1.17).[29]

The solar method used to generate the large amounts of H_2 necessary to address the energy-intensive needs of modern society, on the other hand, relies on CSP and is a catalytic thermochemical process that makes use of concentrated solar radiation to create, from water and sunlight (Figure 1.18), a large surplus of hydrogen suitable for massive energy generation.

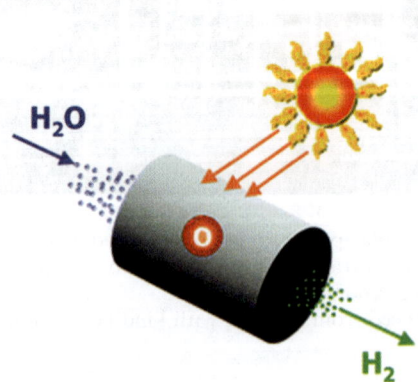

Figure 1.18 Thermochemical water splitting is a catalytic process that makes use of concentrated solar radiation to create a large surplus of hydrogen suitable for energy-intensive applications.

1.4   Hydrogen Safety and Sustainability

Given that it is so light, in the early 1900s hydrogen was used for about two decades to lift huge dirigibles until, on 6 May 1937, the *Hindenburg*, a rigid airship lifted by hydrogen, caught fire in midair over New Jersey and was destroyed during its attempt to dock with its mooring tower at the Lakehurst Naval Air Station (Figure 1.19). Out of 97 people aboard, 62 survived the crash at Lakehurst, although many suffered serious injuries, but 35 passengers perished, along with one member of the civilian landing party.

Ignition of leaking hydrogen is widely assumed to have been the cause of this disaster, even if the visible flames were from combustion of the aluminized fabric coating.[30] Whatever the cause, the accident was broadcast live on radio and filmed so that permanent damage was done to the public reputation of the use of hydrogen in the transport industry. Almost a century later, however, in the same country, and according to a 2010 Pew Research Center poll (Figure 1.20) by 78% *vs.* 17%, the public wants to see increased federal funding for research on wind, solar, and hydrogen technology.[31]

For decades, in any case, the space launch industry has been the only branch of industry to utilize hydrogen fuel in large quantities. Alas, a serious incident was also experienced in this case, namely the Space Shuttle *Challenger* disaster of 1986, when a leak in the liquid hydrogen tank, located in the aft portion of the external tank, caused the explosion of the shuttle.[32]

Figure 1.19   The Zeppelin LZ 129 *Hindenburg* catching fire on 6 May 1937 at Lakehurst Naval Air Station in New Jersey.
(Reproduced from Wikipedia.org, with kind permission.)

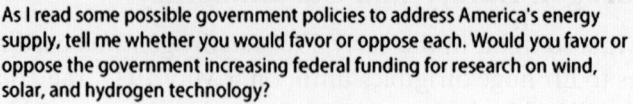

As I read some possible government policies to address America's energy supply, tell me whether you would favor or oppose each. Would you favor or oppose the government increasing federal funding for research on wind, solar, and hydrogen technology?

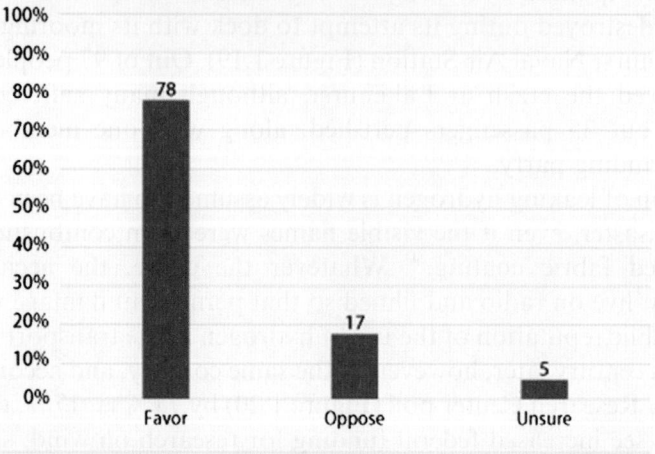

Source: February 3–9, 2010 Pew Research Center for People and the Press political survey.
N=705.

Figure 1.20   A 2010 poll in the USA clearly shows that the public supports climate and energy legislation.
(Image courtesy of Pew Research Center, reproduced from Ref. 31, with kind permission.)

Hydrogen has one of the widest ranges for the explosive/ignition mixture with air of all the gases. This means that, whatever the mix proportion between air and hydrogen, a hydrogen leak will most likely lead to an *explosion*, not a mere flame, when a flame or spark ignites the mixture. This makes the use of hydrogen particularly dangerous in enclosed areas such as tunnels or underground parking.[33]

Hydrogen is odorless and leaks cannot be detected by smell. Moreover, pure hydrogen–oxygen flames burn in the ultraviolet color range and are nearly invisible to the naked eye, so one of the main measures required to implement higher safety standards is early leak detection with hydrogen sensors as well as a flame detector to detect whether a hydrogen leak is burning.

On the other hand, hydrogen is buoyant in air (14.4 times lighter than air, rising at a rate of $20\,\mathrm{m\,s^{-1}}$), thus the gas quickly dilutes and disperses. If released accidentally, hydrogen will rapidly escape upwards (H_2 has a diffusivity in air 3.8 times faster and rises 6 times faster than natural gas). Like gasoline, hydrogen is highly flammable. Yet, as Figure 1.21 shows, owing to the buoyancy of hydrogen the flame shoots up

Figure 1.21 On the left is a vehicle with a hydrogen tank, and on the right a vehicle with a standard gasoline tank. Both tanks have been deliberately punctured and ignited. The top panel shows the two vehicles three seconds after ignition. The bottom panel shows the two vehicles 60 seconds after ignition. (Photo courtesy of the University of Miami, reproduced from Physorg.com.)

vertically, whereas gasoline is heavy and spreads beneath the vehicle. Hence, 60 seconds from ignition, while the hydrogen supply has burned off and the flame is diminished, the gasoline fire has accelerated and has totally engulfed the vehicle on the right.

Another relevant fact is that, during the combustion of hydrogen, the risk of secondary fires is greatly reduced. For example, a flame of 9% hydrogen in air does not ignite a paper sheet even after 60 seconds of exposure.[34] Indeed, the combustion reaction produces water vapor, and this means that only one-tenth of the radiant heat of a hydrocarbon fire is produced.

Hydrogen tanks have even been tested (at the Lawrence Livermore Laboratories in the USA) under extreme conditions, including firearm shots and major mechanical damage and being subjected to flames at 1000 °C for over an hour, without incident (Figure 1.22).[35]

Indeed, the 100-passenger hydrogen-powered fuel cell boat *Alsterwasser* (Figure 1.23), which operates on Alster Lake and the River Elbe in Hamburg, Germany, caught fire in spring 2010, destroying much of the ship. No one was injured and the two 50 kW fuel cells powering a 100 kW hybrid electric propulsion system survived the fire intact and did not even need replacing.[36] All emergency systems operated well, with no leakage or heating of the hydrogen tanks and an immediate automatic shutdown of the fuel cell.

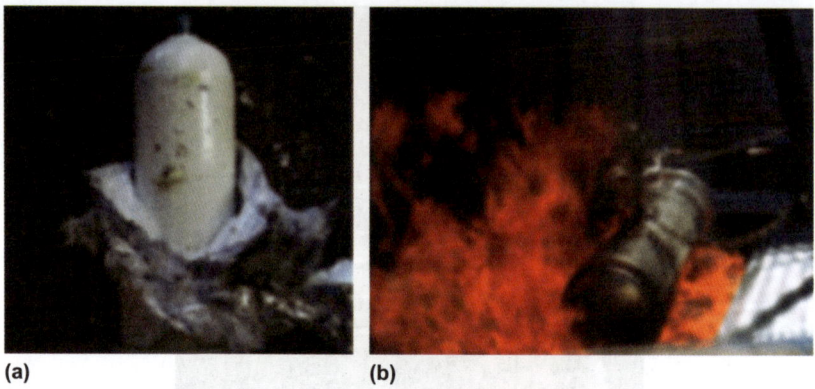

(a)   (b)

Figure 1.22   Destructive tests on full hydrogen tanks. Under extreme conditions (a) piercing the tank with .30-caliber armor-piercing bullets, and (b) bathing the tank in flames for over 60 minutes at 1000 °C, no explosion occurred. (Photo courtesy of the Lawrence Livermore Laboratories, reproduced from Ref. 35, with kind permission.)

Figure 1.23   The *Alsterwasser* fuel cell ship operating near Hamburg, Germany, caught fire in spring 2010, destroying much of the ship. No one was injured and the two 50 kW fuel cells survived the fire intact.
(Photo courtesy of Alster-Touristik GmbH, reproduced from Ref. 28, with kind permission.)

The fire on board was caused by a fault in the connection of the lead acid batteries, which overheated. Remarkably, the *Alsterwasser* is no longer considered a test application and has been operating daily as a ferry in normal service in the Alster-Touristik (ATG) fleet since mid 2011, without any technical problems. As such it no longer requires special maintenance and support and ATG has signed a regular service contract for the fuel cell system with Proton Motor Fuel Cell.

According to both the Canadian[37] and German governments, hydrogen fueling and handling is as safe as, or safer than, compressed natural gas (CNG) fueling. In Germany, for example, there are no additional safety guidelines beyond those for all other combustible gases. According to Germany's 1993 Federal Model Garage Ordinance, there are no usage restrictions for gas-operated vehicles, and hydrogen-powered cars are allowed to drive into parking lots and private garages.[38] In general, it is expected that the general public will be able to use hydrogen technologies in everyday life with at least the same level of safety and comfort as with today's natural gas.

Since 2005, the International Association for Hydrogen Safety (IA-HySafe)[39] has organized an international conference on hydrogen safety to improve public awareness and trust in hydrogen technologies by communicating a better understanding of both the hazards and the risks associated with hydrogen and their management. A useful tool to encourage and promote the safe use of hydrogen is the *Hydrogen Safety Best Practices Manual*,[40] developed in the USA by the Pacific Northwest National Laboratory.

Organized into three major sections (Table 1.5), the online manual captures a vast knowledge base of hydrogen experience from many references and resources into an easy-to-use web-based manual and makes it publicly available, providing suggestions and recommendations on the safe handling and use of hydrogen in various settings.[41] The digital and interactive nature of the tool makes it easy to add new content with assistance from industry experts, such as the recent new

Table 1.5 The major sections in the *Hydrogen Safety Best Practices Manual*. (Reproduced from Ref. 40, with kind permission.)

Section	Contents
Hydrogen Properties	Focused mainly on hydrogen combustion and liquid hydrogen expansion
	Compares key properties of hydrogen and commonly used fuels
Safety Practices (management-oriented)	Safety Culture
	Safety Planning
	Incident Procedures
	Communications
Design and Operations (engineering-oriented)	Facility Design Considerations
	Storage and Piping
	Operating Procedures
	Equipment Maintenance
	Laboratory Safety

section on indoor refueling of hydrogen-powered industrial trucks, as well as noting feedback from users. The European Commission, in its turn, has funded the higher education *HySAFER* program in hydrogen safety engineering at the University of Ulster,[42] building on the European consortium *HySafe: Safety of Hydrogen as an Energy Carrier.*[43]

From a standpoint of "triple bottom line" sustainability (Figure 1.24), solar hydrogen is environmentally as well as economically and societally sustainable.

It is environmentally sustainable because solar hydrogen comes from water and is recombined with air to yield water again, closing the cycle of energy and raw materials with practically no impact on the Earth's atmosphere, and thus on climate change. As molecular hydrogen leaks slowly from most containment vessels, there have been some concerns over the possible environmental impact of a hydrogen economy on the stratosphere problems related to hydrogen gas leakage.[44] However, at most such leakage is likely to be no more than 1–2%, even with widespread hydrogen use and with 2003 technology.[45] The leakage rate today, for example in Germany, is only 0.1% (less than the natural gas leakage rate of 0.7%).

Economically, solar hydrogen is sustainable because both water and primary solar energy are ubiquitous, affording a renewable fuel at low marginal cost that can be used to power the energy needs of both families and enterprises, and allowing the costly world energy trade system based on hydrocarbons and coal to cease. Hydrogen will be produced instead of fossil fuels and mainly distributed through hydrogen pipelines, far shorter and more economical than those connecting the oil-rich countries in Africa and Asia to the developed countries in Europe and in Asia itself.

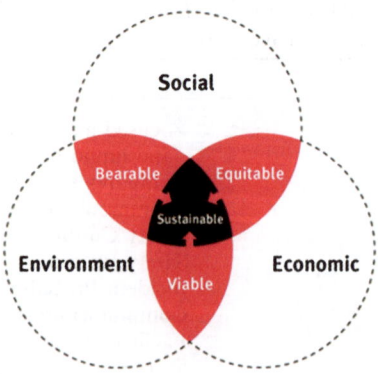

Figure 1.24   Social, economic, and environmental factors are generally involved in sustainability problem solving and decision making.
(Image reproduced from parse.howdesign.com, with kind permission.)

From a social viewpoint, finally, the switch to a solar hydrogen economy will be beneficial because on the one hand it will create the millions of new jobs needed to develop the new energy infrastructure and also because it will free up the immense resources in rich countries that are currently used to purchase oil, gas and coal. Taxation will reduce and financial resources will instead be employed to clean up the damage created by 2.5 centuries of the massive utilization of fossil fuels.

1.5 Hydrogen as Energy Carrier: Exergizing the Energy System

Recently, Joshi and co-workers evaluated the performance of the two routes of hydrogen production that are relevant to this textbook, namely PV and solar thermal systems, based on exergy analysis and a "sustainability index".[46] As expected, the solar thermal hydrogen production system has a higher sustainability index because of higher exergy efficiency, when compared with the PV hydrogen production system whose exergy efficiency ranges between 3.68 and 4.84%. The exergy efficiencies of concentrating the collector increase with increasing solar radiation.

In brief, whenever energy is converted (produced, handled, stored, transported, disseminated, utilized, *etc.*) it is split into two parts (Equation 1.4):

$$Energy = exergy + anergy \qquad (1.4)$$

Exergy is the maximum amount of available technical work extracted from energy (Figure 1.25). It can be converted into any other form of energy, whereas anergy cannot. Hydrogen-supplied fuel cells are highly efficient exergetically, generating electricity (= pure exergy), whereas the exergy content of the remaining heat at the fuel cell's specific temperature can be used to meet the requirements of industrial or residential heating.

Hydrogen exergizes the energy system because novel solar-electric systems, such as hydrogen-supplied fuel cells or solar photovoltaics, are *not* Carnotian heat engines. In other words, hydrogen minimizes anergy, making more technical work available (exergy) out of energy.

First conceived over 200 years ago, classical Carnotian energy systems – combustion-based engines and boilers – are obsolete and poorly efficient but still widely employed in power plants, in residential or industrial heating systems, and in the engines of automobiles, trucks and airplanes. They produce too much heat in the wrong place where no potential user asks for it. For example, central heating system boilers

Figure 1.25   A visual analogy explaining the concept of exergy in terms of the
                 toothpaste (energy) made available by squeezing the tube. Entropy, in
                 this analogy, is the depression in the tube.
                 (Image courtesy of Dr F. Günther, reproduced from www.holon.se/
                 folke, with kind permission.)

(Figure 1.26) are, to paraphrase Winter again, "exergetically
miserable".[47]

   They energetically convert almost 100% of the chemical energy of the
fuel into heat. Absurd as it may seem, we generate 1000 °C combustion
temperatures simply in order to provide a room radiator temperature of
some 60 to 70 °C.

   Hydrogen-fueled fuel cells, however, generate electricity (= pure
exergy) efficiently firsthand, and the leftover heat still suffices to heat a
house for most of the year. In the words of Balzani:[48]

> Electricity is the most flexible and convenient form of energy. It can
> be converted without any losses to useful heat, even to generate
> temperatures higher than combustion of any fossil fuel; it can be
> turned with high efficiency into mechanical energy; it is used to
> create light; it can be easily adjusted with very high precision and is
> clean and silent at the point of consumption.
>
> Today electricity is used to generate any kind of artificial lighting,
> to heat and cool buildings, to run our information communication
> technology, to propel trains and buses, to supply factory machinery
> with reliable and inexpensive energy, to create entirely new
> industries, and to perform a variety of other useful achievements

Figure 1.26 A central heating system boiler typically burns natural gas (1000 °C) to produce hot water at 60–70 °C.
(Photo courtesy of the University of Miami, reproduced from parse. howdesign.com.)

that make our life cleaner, easier, more secure and enjoyable. In the future, electricity is also expected to power millions of personal vehicles. There are indeed good reasons to believe that we are going towards an electricity-based economy.

Today, hydrogen-fueled fuel cells, originally developed in the 1960s by the US space agency (NASA) for space programs to provide power *and* water for space flights, are compact energy devices whose module energy output ranges over 6 orders of magnitude from watts to a few megawatts.[49]

Fuel cells are thus widely used for a variety of different applications including stationary power sources, materials handling vehicles (forklifts), and recharging small portable electronics (Figure 1.27).

Hydrogen fuel cells are highly efficient, with low noise, without moving parts and, thus, with no major dynamic forces or moments. They are being developed for portable electronics, for mobile on-board delivery of electricity to the electrical on-board grids and the drive trains

Figure 1.27   Fuel cells are today a commercial reality.
(Image courtesy of Cleantech Investor.)

of automobiles, buses, trucks, locomotives, and most recently also for airplanes. Apple, for example, the maker of the *iPhone*, is currently considering powering its successful portable electronic devices with a hydrogen fuel cell.[50]

Perhaps not surprisingly, therefore, the fuel cell industry is growing even in the face of the global economic recession that started in 2008, with 2010 revenues exceeding US$750 million.[51] Markets such as uninterruptible power supplies (UPS), residential combined heat and power (CHP), power for remote monitoring equipment, auxiliary power units (APUs), and portable power for military applications are all experiencing a rapid increase. In this context of progress and growth, hydrogen is clearly emerging as an energy storage medium.

1.6   The Hydrogen Science & Technology Network

In 1974, a few scholars in the United States started the International Association for Hydrogen Energy (IAHE).[52] In 1981 they established the *International Journal of Hydrogen Energy*, published monthly by Elsevier since 1982. The journal is nowadays one of the primary literature sources for the exchange and dissemination of basic ideas in the field of hydrogen energy, whereas the World Hydrogen Energy Conference, organized by the IAHE every two years on a different continent, attracts about 1500 attendees. The monthly *Hydrogen & Fuel Cell Letter* (www.hfcletter.com), which covers the science, business and politics of hydrogen and fuel cells, has been published continuously since 1986 and is widely regarded as an important voice in the international hydrogen community.

From Germany to Japan, from the USA to Korea and Australia, there is a growing number of national and international organizations devoted to hydrogen energy and its applications. In the USA, the Fuel Cell and Hydrogen Energy Association (FCHEA)[53] is a primary advocacy organization dedicated to the commercialization of fuel cells and hydrogen energy technologies. Its membership represents the full spectrum of the supply chain from universities, government laboratories and agencies, to trade associations, fuel cell manufacturers and hydrogen producers.

The US Government Department of Energy (DOE) operates a "Hydrogen and Fuel Cells Program" whose website[54] offers a portal to information about the Department's research and development in hydrogen production, delivery, storage, and utilization, as well as activities in technology validation, safety standards and education. In the USA, so advanced is hydrogen technology that in September 2011 Air Products in California opened a fuelling station that draws hydrogen from the methane gas created while wastewater sits in holding tanks (Figure 1.28).[55]

In Europe, the Fuel Cells and Hydrogen Joint Undertaking (Table 1.6)[56] is a public–private partnership whose three members are the European Commission, fuel cell and hydrogen industries (represented by the NEW Industry Grouping) and the research community (represented by the Research Grouping N.ERGHY). The organization supports research, technological development and demonstration activities in fuel cell and hydrogen energy technologies, aiming to accelerate the market introduction of these technologies through a concentrated effort from all sectors.

On the same continent, the European Hydrogen Association (EHA)[57] currently represents 20 national hydrogen and fuel cell organizations. Since 2004, the European Commission has extensively funded the HyFLEET:CUTE project to promote hydrogen-fueled buses in eight European cities, but also in Beijing and Perth.[58] The project has clearly

Figure 1.28   At this refuelling station in the USA, hydrogen is generated from a municipal wastewater water treatment facility in Orange County Sanitation District.
(Photo courtesy of Air Products; reproduced from Ref. 55, with kind permission.)

shown that today's hydrogen technology allows buses to run efficiently and cleanly on hydrogen, provided that the cost of the fuel cells (and thus of the buses) is significantly reduced.[59]

Between 2004 and 2009, the project has successfully operated more than 40 hydrogen-powered buses over four years in 10 cities on three continents, with over 8.5 million people transported in regular public transport operations; more than 2.5 million km has been travelled, and in excess of 550 000 kg of H_2 safely refueled. The project has shown that the infrastructure to produce, supply and distribute hydrogen for transport can be implemented efficiently and safely by generating H_2 on-site using renewable energy (Figure 1.29).

In Japan, between 2002 and 2010, the Ministry of Economy, Trade and Industry (METI) funded the "Japan Hydrogen & Fuel Cell Demonstration Project",[60] which aimed to advance both fuel cell vehicles (FCVs) and the hydrogen energy infrastructure for full-scale mass production and dissemination of hydrogen-powered cars and vans. Japanese partners collected basic data on hydrogen generation, FCV performance, environmental features, energy efficiency and safety.

In Russia, the world's largest country and the world's largest producer of oil (production is around 10 120 000 barrels per day),[61] renewable energy technologies are underdeveloped. Yet, Russia had an

Table 1.6 A selection of industry and research institute partners in the European FCH-JU. (Adapted from Ref. 56, with kind permission.)

Industrial companies	*Research institutions*
United Kingdom	**United Kingdom**
Adelan	UK Energy Research Centre
AFC Energy	
Alstom	
Air Products	
Johnson Matthey	
Intelligent Energy	
Rolls Royce Fuel cell system	
Germany	**Germany**
Adam Opel GmbH	DLR – Deutsches Zentrum für Luft – und
Daimler	Raumfahrt
EWE AG	FZJ – Forschungszentrum Jülich
Linde	FZK – Forschungszentrum Karlsruhe
NuCellSys	ZSW – Zentrum für Sonnenenergie – und
Umicore AG	Wasserstoffforschung
Vattenfall	
Volkswagen	
France	**Greece**
Air Liquide	CPERI – Chemical Process Engineering
Saint Gobain Centre de	Research Institute
Recherches et d'Etudes	CRES – Centre for Renewable Energy Sources
Européen	
SNECMA	
Total France	
Italy	**Spain**
Ansaldo Fuel Cells	CIDETEC – Centro de Tecnologías
Electro Power Systems	Electroquímicas
Enel Produzione	CIEMAT – Centro de Investigaciones
Environment Park	Energéticas
HySyTech S.r.l.	CNH2 – The National Centre for Hydrogen
SOFCpower	and Fuel Cell Technology
	INTA – Instituto nacional de Técnica
	Aeroespacials
	University of Alicante

enormous advantage in the early 2000s in hydrogen technology.[62] In the 1980s Russian scientists created the first hydrogen-driven plane, a Tu-155 aircraft (Figure 1.30);[63] the Russian space shuttle "Buran" and some Russian submarines also had liquid hydrogen as a power source.

However, development then stopped because, while Japan, America, and Europe invested billions of dollars, oil- and gas-rich Russia

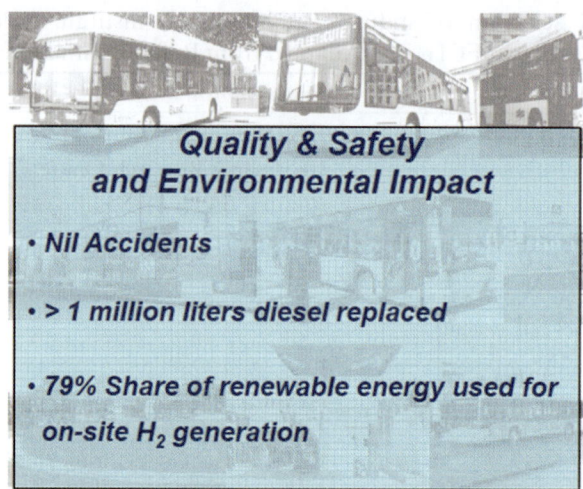

Figure 1.29 Quality, safety and environmental impact of the HyFLEET:CUTE project as of September 2009.
(Reproduced from Ref. 59, with kind permission.)

Figure 1.30 On 15 April 1988, in Russia, this Tupolev Tu-155 aircraft performed its maiden flight using liquid hydrogen. Cryogenic fuel was kept in a fuel tank of 17.5 m³ capacity installed in a special compartment in the rear portion of the passenger cabin.
(Photo courtesy of Tupolev, reproduced from Ref. 63, with kind permission.)

allocated just a few million rubles. In the words of the president of Russia's National Association of Hydrogen Energy:

Once I went to Anatoly Chubais [NE: former head of the dismantled Russian electricity monopoly and present head of

RUSNANO Corporation], my friend and colleague from St-Petersburg, asking him for support for the hydrogen energy project.

He told me that a kilowatt of installed hydrogen capacity cost US$3,000, when actually it was US$5,000, and then said: "Come again when the price is US$100". During the following 5 years, the price of the hydrogen technologies got reduced to $3,000, while during the same period, the price of 1 kW of installed electric capacity rose to US$2,500 in Moscow and St-Petersburg, I would never imagine it would raise so much on his side! 500US$ is a considerable difference but not insurmountable with a bit more time...

Today, almost silently, hydrogen motorization is rapidly approaching.[64] The third annual assessment by *H2stations.org*[65] found that 22 new hydrogen refueling stations opened worldwide in 2010, increasing the total number to 212 (Figure 1.31). About 80 fueling stations are in the USA, 27 in Germany, 6 in China, 4 in Italy and none in Russia.

Throughout Europe, 11 new hydrogen refueling stations began operation in 2010, while 5 new stations opened in the USA and another 9 will be completed shortly. However, some refueling stations closed, so the database shows the number of hydrogen refueling stations in operation to be 80 each in Europe and the USA and 48 in Asia. Another 127 refueling stations are in the planning stage around the world.

Updated on an ongoing basis (and free for non-commercial users) the *H2stations.org* online database includes detailed information about 418 refueling stations that are either already in existence or for which planning is underway. For example, in Germany the automotive manufacturer Daimler AG and the Linde technology group plan to open 20 new filling stations to enable every location in Germany to be reached by a fuel-cell vehicle by the year 2015. In 2009, eight automakers (Ford Daimler AG, Ford Motor Company, General Motors, Honda, Hyundai Motor Company, Kia Motors Corporation, Renault/Nissan, and Toyota) signed an agreement to bring FCVs to the market by 2015.

In general, the current major players in hydrogen fueling are the industrial gas companies, and the energy and gas companies (Table 1.7). Smaller "independent" hydrogen suppliers that are developing and marketing smaller onsite hydrogen generator technologies could offer a more modular path to hydrogen infrastructure buildout.

A market report[66] recently forecasted that more than 5200 hydrogen fueling stations for cars, buses and forklifts will be operational worldwide by 2020, up from just 200 stations in 2010. The same report

Figure 1.31   Map of the existing hydrogen filling stations worldwide as of April 2011, according to *H2stations.org*.
(Images courtesy of TÜV SÜD and Ludwig-Bölkow-Systemtechnik, reproduced from H2stations.org, with kind permission.)

Table 1.7 Key industry players with hydrogen fueling stations for cars, buses and forklifts. (Adapted from Ref. 66, with kind permission.)

Industrial gas companies	Independent infrastructure developers
Air Liquide	H2 Logic
Air Products	Hydrogenics
Linde	ITM Power
Praxair	Nuvera
	Proton Onsite and SunHydro

anticipates that the increased utilization of hydrogen as a fuel will drive annual demand from approximately 775 tonnes in 2010 to 418 000 tonnes by 2020.

References

1. J. S. Rigden, *Hydrogen: The Essential Element*, Harvard University Press, Cambridge, MA, 2003.
2. A. Stwertka, *A Guide to the Elements*, Oxford University Press, Oxford, 1996.
3. M. Fichtner and F. Idrissova, in *The Hydrogen Economy: Opportunities and Challenges*, ed. M. Ball and M. Wietschel, Cambridge University Press, Cambridge, UK, 2009, p. 271.
4. A. Züttel, Materials for hydrogen storage, in *Catalysis for Sustainable Energy Production*, ed. C. Bianchini and P. Barbaro, Wiley-VCH, Weinheim, 2009, p 235.
5. W. B. Leung, N. H. March and H. Motz, *Phys. Lett. A*, 1976, **6**, 425.
6. The lower heating of a fuel is defined as the amount of heat released by combusting a specified quantity (initially at 25 °C) and returning the temperature of the combustion products to 150 °C, which assumes that the latent heat of vaporization of water in the reaction products is not recovered.
7. US Department of Energy, Fuel Cell Technologies Program, Hydrogen Properties, in *Fuel Cell Engines and Related Technologies*, Rev 0, December 2001. www1.eere.energy.gov/hydrogenandfuelcells/tech_validation/pdfs/fcm01r0.pdfGT2gymcozjg (last accessed on 03/01/2012).
8. N. Armaroli and V. Balzani, *ChemSusChem*, 2011, **4**, 21.
9. H. C. Urey, F. G. Brickwedde and G. M. Murphy, *Phys. Rev.*, 1932, **40**, 1.
10. For a splendid historical account on this topic, see A. Pais, *Inward Bound: Of Matter and Forces in the Physical World*, Oxford University Press, Oxford, UK, 1988.

11. V. I. Tikhonov and A. A. Volkov, *Science*, 2002, **296**, 2363.
12. Y. Y. Milenko, R. M. Sibileva and M. A. Strzhemechny, *J. Low Temp. Phys.*, 1997, **107**, 77.
13. R. E. Svadlenak and A. B. Scott, *J. Am. Chem. Soc.*, 1957, **79**, 5385.
14. C. G. Sluijter, H. F. P. Knaap and J. J. M. Beenakker, *Physica*, 1964, **30**, 745.
15. R. Berman, A. H. Cooke and R. W. Hill, *Ann. Rev. Phys. Chem.*, 1956, **7**, 1.
16. W. Grove, *Philos. Magazine J. Sci.*, 1842, **XXI**, 417.
17. BCC Research, *Hydrogen as a Chemical Constituent and as an Energy Source*, Report Code: CHM031C, Wellesley, MA, 2011.
18. M. A. Bernal, L. Pampín, C. Andrade and P. M. Bello Bugallo, Beyond the simplicity: optimizing the hydrogen production process, in *World Renewable Energy Congress*, ed. B. Moshfegh, Linköping, Sweden, 8–13 May, 2011, Linköping Electronic Conference Proceedings, 57, Linköping University Electronic Press. http://dx.doi.org/10.3384/ecp11057 (last accessed on 03/01/2012).
19. C. Koroneos, A. Dompros, G. Roumbas and N. Moussiopoulos, *Int. J. Hydrogen Ener.*, 2004, **29**, 1443.
20. www1.eere.energy.gov/vehiclesandfuels/facts/favorites/fcvt_fotw205.html (last accessed on 02/01/2012).
21. Pike Research, *Fuel Cell Vehicles*, Report, Boulder, CO, April 2011. www.pikeresearch.com/research/fuel-cell-vehicles (last accessed on 03/01/2012).
22. G. W. Bush, State of the Union Address (January 31, 2006). Available at: www.whitehouse.gov/stateoftheunion.
23. C.-J. Winter, *Int. J. Hydrogen Ener.*, 2009, **34**, S1.
24. J. J. Romm, *The Hype about Hydrogen: Fact and Fiction in the Race to Save the Climate*, Island Press, Washington, DC, 2004.
25. V. Smil, 21st century energy: Some sobering thoughts, *OECD Observer*, 2006, **258/59**, 22.
26. T. Bradford, *Solar Revolution: The Economic Transformation of the Global Energy Industry*, MIT Press, Boston, MA, 2006.
27. R. L. Hirsch, R. Bezdek and R. Wendling, *Peaking of World Oil Production: Impacts, Mitigation and Risk Management*, February 2005, Report by Robert L. Hirsch *et al.*, commissioned by the US Department of Energy. Available at: www.netl.doe.gov/energy-analyses/pubs/Oil_Peaking_NETL.pdf (last accessed on 03/01/2012).
28. J. Wise, The truth about hydrogen, *Popular Mechanics*, November 1, 2006, www.popularmechanics.com/science/energy/next-generation/4199381 (last accessed on 02/01/2012).

29. http://world.honda.com/FuelCell/FCX/station/ (last accessed on 02/01/2012).
30. A. J. Dessler, D. E. Overs and W. H. Appleby, The Hindenburg fire: hydrogen or incendiary paint? *Buoyant Flight*, 2005, **52**, no. 2 and no. 3. http://spot.colorado.edu/~dziadeck/zf/LZ129fire2005jan12.pdf (last accessed on 02/01/2012).
31. Memo to policymakers: Public STILL favors the transition to clean energy, 18 March 2010. http://thinkprogress.org/romm/2010/03/18/205647/memo-to-policymakers-public-still-favors-the-transition-to-clean-energy/ (last accessed on 06/02/2012).
32. http://en.wikipedia.org/wiki/Space_Shuttle_Challenger_disaster (last accessed on 02/01/2012).
33. V. P. Utgikar and T. Thiesen, *Technol. Soc.*, 2005, **27**, 315.
34. M. R. Swain, P. A. Filoso and M. N. Swain, *Int. J. Hydrogen Ener.*, 2007, **32**, 287.
35. L. Powers, Flexibly fueled storage tank brings hydrogen-powered cars closer to reality, *Sci. Technol. Rev.*, 2003, 24–26. https://www.llnl.gov/str/June03/Aceves.html (last accessed on 02/01/2012).
36. Proton Power announces service contract for fuel cell ferry, 24 November 2011, www.fuelcelltoday.com/news-events/news-archive/2011/november/proton-power-announces-service-contract-for-fuel-cell-ferry (last accessed on 02/01/2012).
37. I. Mac Intyre, A. V. Tchouvelev, D. R. Hay, J. Wong, J. Grant and P. Benard, *Int. J. Hydrogen Ener.*, 2007, **32**, 2134.
38. www.zukunftsprojektwasserstoff.at/typo/fileadmin/user_upload/download/FAQ_fuel_cell.pdf (last accessed on 02/01/2012).
39. www.hysafe.org/IAHySafe (last accessed on 02/01/2012).
40. *Hydrogen Safety Best Practices Manual*, Pacific Northwest National Laboratory, USA. www.H2BestPractices.org (last accessed on 02/01/2012).
41. L. L. Fassbender (Pacific Northwest National Laboratory), *Hydrogen Safety Knowledge Tools*, 2011, www.hydrogensafety.info/2011/jan/knowledgeTools.asp (last accessed on 02/01/2012).
42. http://hysafer.ulster.ac.uk (last accessed on 02/01/2012).
43. www.hysafe.org/IAHySafe (last accessed on 02/01/2012).
44. T. K. Tromp, R.-L. Shia, M. Allen, J. M. Eiler and Y. L. Yung, *Science*, 2003, **300**, 1740.
45. D. M. Kammen and T. E. Lipman, *Science*, 2003, **302**, 226.
46. A. S. Joshi, I. Dincer and B. V. Reddy, *Int. J. Hydrogen Ener.*, 2010, **35**, 4901.

47. C.-J. Winter, *Energy Policy is Technology Politics – The Hydrogen Energy Case*, www.itshytime.de/hytime/Kreta1-a.pdf (last accessed on 02/01/2012).
48. N. Armaroli and V. Balzani, *Energy Environ. Sci.*, 2011, **4**, 3193.
49. D. Stolten, *Hydrogen and Fuel Cells*, Wiley-VCH, Weinheim, 2010.
50. Application 20110311895, filed in August 2010, covers a fuel cell system for a portable computing device, including a fuel source, a controller and an interface to the device. Victoria Slind-Flor, Could next Apple iPhone, iPod be powered by a hydrogen fuel cell?, *Washington Post*, 27 December 2011. See also at: www.washingtonpost.com/business/could-next-apple-iphoneipod-be-powered-by-a-hydrogen-fuel-cell/2011/12/27/gIQAN3DPMP_story.html (last accessed on 02/01/2012). Apple, based in Cupertino, California, acknowledged in the application that "it is extremely challenging to design hydrogen fuel cell systems which are sufficiently portable and cost-effective to be used with portable electronic devices."
51. Pike Research, *Fuel Cell and Hydrogen Industry: Ten Trends to Watch in 2011 and Beyond*, White Paper, Boulder, CO, February 2011. www.pikeresearch.com/research/fuel-cell-and-hydrogen-industryten-trends-to-watch-in-2011-and-beyond (last accessed on 02/01/2012).
52. T. Nejat Veziroglu, *Int. J. Hydrogen Ener.*, 2000, **25**, 1143.
53. www.fchea.org (last accessed on 02/01/2012).
54. www.hydrogen.energy.gov (last accessed on 02/01/2012).
55. E. Loveday, World's first wastewater-to-hydrogen fueling station opens in California, *Green Car News*, September 2, 2011. www.green.autoblog.com/2011/09/02/air-products-opens-worlds-first-wastewater-to-hydrogen-fueling/ (last accessed on 02/01/2012).
56. www.fch-ju.eu (last accessed on 02/01/2012).
57. www.h2euro.org (last accessed on 02/01/2012).
58. M. Kentzler (Daimler AG), *HyFLEET:CUTE. The Achievements of the World's Largest Hydrogen Powered Bus Fleet*, HyFLEET:CUTE Conference, Hamburg, Germany, 17 November 2009. http://hyfleetcute.com/data/Kentzler_AchievementsWorldLargestH2Fleet.pdf (last accessed on 02/01/2012).
59. *Hydrogen Transport – Bus technology and fuel today for a sustainable future*, press release, HyFLEET:CUTE Conference, Hamburg, Germany, 17 November 2009. www.global-hydrogen-bus-platform.com/news/item/100/ (last accessed on 02/01/2012).
60. www.jari.or.jp/jhfc/e/jhfc/index.html (last accessed on 02/01/2012).

61. Russia overtakes Saudi Arabia as world's biggest crude oil producer, *Euromonitor Int.*, September 28, 2010, http://blog.euromonitor.com/ 2010/09/russia-has-overtaken-saudi-arabia-as-worlds-biggest-crude-oil-producer.html (last accessed on 02/01/2012).
62. Interview with Petr Shelisch, President, Russia National Association of Hydrogen Energy, www.russiaenergy.com/index.php# state=InterviewDetail&id=456 (last accessed on 02/01/2012).
63. www.tupolev.ru/english/Show.asp?SectionID=82 (last accessed on 02/01/2012).
64. J. Motavalli, Questions for Peter Hoffmann: A hydrogen advocate whose time may have come, *The New York Times,* 2 February 2012, www.nytimes.com/2012/02/02/automobiles/wheels/a-hydrogen-advocate-whose-time-has-come.html (last accessed on 06/02/2012).
65. A recently established Internet-based database of TÜV SÜD and Ludwig-Bölkow-Systemtechnik, a German energy and environmental consultancy, http://www.h2stations.org/ (last accessed 02/04/ 2012).
66. Pike Research, *Hydrogen Infrastructure*, Report, Boulder, CO, July 2011. www.pikeresearch.com/research/hydrogen-infrastructure (last accessed on 02/01/2012).

CHAPTER 2

Water Electrolysis with Solar Electricity

2.1 Water Electrolysis

Electrolytic production of hydrogen by water electrolysis is a well known electrochemical process first described by William Nicholson and Anthony Carlisle in 1800, a few *weeks* after Volta described his new "voltaic" pile.[1] In detail, the splitting of one mole of water into gaseous H_2 and O_2 by the action of electricity (Equation 2.1, where F is the Faraday constant measuring 1 mole of electricity, 96 485 C) produces a mole of hydrogen gas and a half-mole of oxygen gas in their normal diatomic forms:

$$H_2O(l) + 2F \rightarrow H_2(g) + \tfrac{1}{2}O_2(g) \qquad (2.1)$$

An electrolyzer uses electricity to split water into the component elements. When sufficient (critical) voltage is applied across the electrodes, current will flow and oxygen gas (O_2) will form at the anode and hydrogen gas (H_2) at the cathode (Figure 2.1).

Figure 2.1 describes the process in terms of the Gibbs free energy as well as the other thermodynamic potentials. Using the relevant values of thermodynamic properties at 298 K and 1 atmosphere pressure (Table 2.1), the process must provide the energy for the dissociation plus the energy to expand the gases produced, affording a change in enthalpy as shown in Table 2.1.

Solar Hydrogen: Fuel of the Future
Mario Pagliaro and Athanasios G. Konstandopoulos
© Mario Pagliaro and Athanasios G. Konstandopoulos 2012
Published by the Royal Society of Chemistry, www.rsc.org

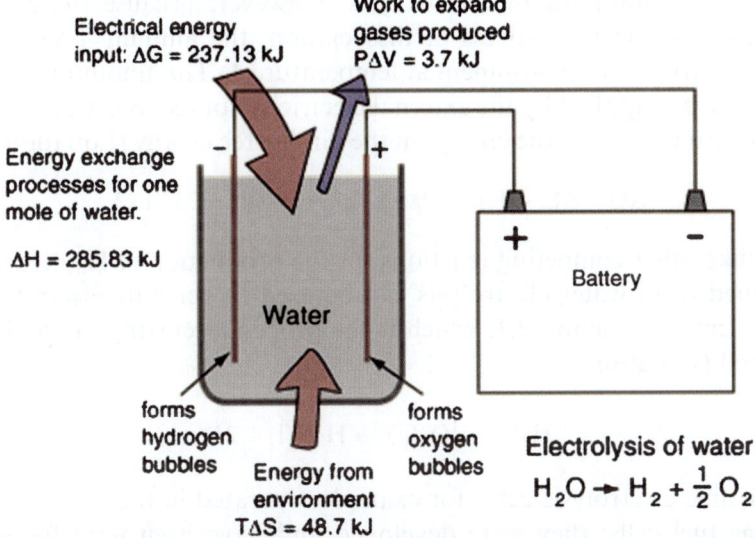

Figure 2.1 A simplified diagram showing the basics of water electrolysis. The wires protruding into the solution are the cathode and anode, where reactions of hydrogen and oxygen evolution take place.
(Reproduced from Ref. 2, with kind permission.)

Table 2.1 Thermodynamic properties of selected substances for one mole at 298 K and 1 atmosphere pressure. (Reproduced from D. V. Schroeder, *An Introduction to Thermal Physics*, Addison Wesley, 2000, with kind permission.)

Thermodynamic potential	H_2O	H_2	$0.5\,O_2$	Change
Enthalpy	−285.83 kJ	0	0	$\Delta H = 285.83$ kJ
Entropy	$69.91\ \mathrm{J\,K^{-1}}$	$130.68\ \mathrm{J\,K^{-1}}$	$0.5 \times 205.14\ \mathrm{J\,K^{-1}}$	$T\Delta S = 48.7$ kJ

At temperature 298 K and 1 atmosphere pressure, the system work is (Equation 2.2):[2]

$$W = P\Delta V = \left(101.3 \times 10^3\,\mathrm{Pa}\right)\left(1.5\,\mathrm{moles}\right)\left(22.4 \times 10^{-3}\mathrm{m}^3/\mathrm{mol}^{-1}\right)$$
$$\left(298\,\mathrm{K}/273\,\mathrm{K}\right) = 3715\,\mathrm{J} \tag{2.2}$$

Given that the enthalpy $H = U + PV$, the change in internal energy U is then (Equation 2.3):

$$\Delta U = \Delta H - P\Delta V = 285.83\,\mathrm{kJ} - 3.72\,\mathrm{kJ} = 282.1\,\mathrm{kJ} \tag{2.3}$$

This change in internal energy must be accompanied by the expansion of the gases produced, so the change in enthalpy represents the necessary energy to accomplish the electrolysis. However, because the entropy increases during the process of dissociation, the amount TΔS can be provided from the environment at temperature T. The amount that must therefore be supplied by the external electricity source to get the process to proceed is actually the change in the Gibbs free energy (Equation 2.4):

$$\Delta G = \Delta H - T\Delta S = 285.83 \, kJ - 48.7 \, kJ = 237.1 \, kJ \qquad (2.4)$$

Unlike other competing reactions for the production of H_2, hydrogen obtained from water electrolysis can be used to generate electricity by the reverse of reaction 2.1, which is the process occurring in an H_2–O_2 fuel cell (Equation 2.5):

$$H_2(g) + \tfrac{1}{2}O_2(g) \rightarrow H_2O(l) + 2F \qquad (2.5)$$

Alkaline electrolysis cells, for example, operated in the reverse direction as fuel cells; they were developed and have been used for space exploration since the 1960s.[3]

Like a battery, the hydrogen fuel cell uses a chemical reaction to provide an external voltage, but differs from a battery in that the fuel is continually supplied in the form of hydrogen and oxygen gas, affording only water and heat as reaction products (Figure 2.2).

Now, the entropy of the gases decreases by 48.7 kJ in the process of combination because the number of water molecules is less than the number of hydrogen and oxygen molecules combining. Given that the *total* entropy will not decrease in the reaction, the excess entropy in the amount TΔS must be expelled to the environment as heat at temperature T. The amount of energy per mole of hydrogen which can be provided as electrical energy is the change in the Gibbs free energy (Equation 2.6):

$$\Delta G = \Delta H - T\Delta S = -285.83 \, kJ + 48.7 \, kJ = -237.1 \, kJ \qquad (2.6)$$

The figures, in this ideal case, show that the fuel energy is converted to electrical energy at an efficiency of 237.1/285.8 × 100% = 83%, which is far greater than the ideal, Carnot's efficiency obtainable by burning the hydrogen and using the heat to power an electricity generator. Overall, in the electrolysis/fuel cell pair where the enthalpy change is 285.8 kJ, a total of 237 kJ of energy is required to drive electrolysis while heat from the environment contributes TΔS = 48.7 kJ to assist in the process. In the opposite reaction, occurring in the fuel cell, a total of 237 kJ of energy is obtained as electric energy, while TΔS = 48.7 kJ is released to the environment.

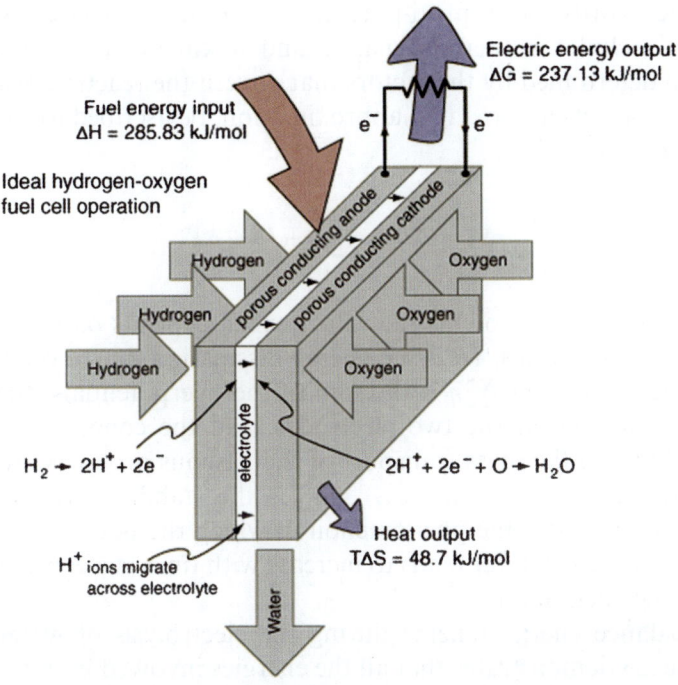

Figure 2.2 A simplified diagram showing the basics of the fuel cell. The porous cathode and anode are typically made of conducting, catalytic materials while an electrolyte ensures ion migration and overall neutrality. (Reproduced from Ref. 2, with kind permission.)

Let us consider again the water dissociation reaction in Equation 2.1 from a thermodynamic viewpoint. The minimum potential difference that can be applied between the electrodes of a water electrolyzer without producing electrolysis is (Equation 2.7):[4]

$$\Delta E = \frac{-\Delta G}{nF}(n=2) \qquad (2.7)$$

Wherein ΔE is the electrode potential under equilibrium conditions. The required voltage, (the standard potential) for the overall reaction (2.1) is equal to 1.23 volts at 25 °C at 1 atmosphere. In reality, more voltage than the standard potential must be applied to get appreciable water splitting and gas production.

In order to drive water electrolysis at a practical rate, a $\Delta V > \Delta E$ must be applied, which implies that part of the electrical energy is used to overcome the reaction resistances (Equation 2.8):

$$\Delta V = \frac{-\Delta G}{nF} + dissipation \qquad (2.8)$$

In other words, the applied potential ΔV to drive the electrochemical cell at I includes a thermodynamic and a kinetic (dissipation) contribution determined by the factors that govern the reaction resistances. Such dissipation can, to a first approximation, be ascribed to three main factors (Equation 2.9):

$$\Delta V = \Delta E + \sum \eta + IR + \Delta V_t \qquad (2.9)$$

where ΔE is the thermodynamic value, which depends on the nature of the electrode reactions; IR is the energy dissipation due to ohmic drops in the electrolytic cell; $\sum \eta$ is the sum of the overpotentials (the activation overpotential at the two electrodes, and the concentration overpotential due to the mass transport of the gaseous products away from the anode and cathode surfaces); ΔV_t is the stability term, namely a factor that expresses the phenomenon in which the potential difference applied to an electrolyzer tends to increase with time as a consequence of performance degradation.[5]

The balance energy scheme during the electrolysis of water shown in Figure 2.3 demonstrates that all the energies involved in this equation of balance, except for ΔA ($\Delta E + \Delta V_t$), will leave the reaction as heat: $T \Delta S$ (Joule heating).

The overpotential η should be kept low in order to maximize the efficiency and to minimize the production of heat. On the other hand, the lower the overpotential the more slowly the reaction will occur, so a compromise is needed. In practice, the activation overpotential increases by increasing the current density and can be lowered by using electrodes that have a catalytic action, such as platinum or palladium. One of the

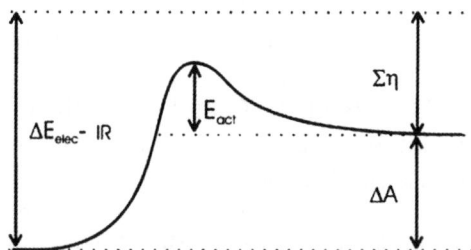

Figure 2.3   Schematic representation of the energies involved in the water electrolysis reaction. To start the reaction, it is necessary to overcome the activation energy E_{act}, the extra energy barrier.
(Reproduced from Ref. 6, with kind permission.)

best ways to increase the current without increasing the overpotential is to increase the contact areas between the electrodes and the liquid.[6]

The consumption of electrical energy in the electrolytic cell where electrolysis is carried out for a time t at a current I with a potential difference ΔV is (Equation 2.10):

$$Energy = \Delta VIt \qquad (2.10)$$

Given that the overall target of any electrochemical process is to minimize the electrical consumption, Equation 2.6 implies that the process should be carried out at the maximum I with the minimum ΔV; given that the current fixes the desired production rate and is more easily controlled, in industrial processes it is the current that is fixed rather than the potential.

2.2 Current Electrolytic Technologies

The kinetics of the O_2 and H_2 evolution reactions at the electrodes limit how fast hydrogen can be generated and are dependent on the electrode's chemical activity. Precious metals such as platinum and palladium generally make good electrodes, but they are prohibitively expensive. The reaction can be "overdriven" by applying a larger voltage than the minimum required, but this reduces the efficiency. For water splitting, the oxygen-evolving anode is the larger contributor to the problem, requiring a larger overpotential.

In general, first, to enhance the water conductivity and thus the overall rate of the process of water electrolysis, an electrolyte is dissolved in water. Three water electrolyser (WE) technologies are available today that are classified according to the electrolyte employed: alkaline (AWE), polymer electrolyte membrane (PEMWE), and solid oxide electrolyzers (SOWE).[7] It is interesting to note that by 1902 more than 400 industrial water electrolysis units were in operation, while the first large water electrolysis plant with a capacity of $10\,000\,Nm^3\,H_2\,h^{-1}$ had started to operate by 1939. In 1966, General Electric built the first solid polymer electrolyte system (SPE), and in 1972 the first solid oxide water electrolysis unit was developed. The first advanced alkaline systems started in 1978. The story ends in recent times with the development of proton exchange membranes, usable for water electrolysis units and fuel cells, by DuPont and other manufacturers. This was a result of the developments in the field of high temperature solid oxide technology and the optimization and reconstruction of alkaline water electrolyzers.[8]

Conventional water electrolysis utilizes an alkaline aqueous electrolyte with a separation membrane, avoiding the remixing of H_2 and O_2; SPE electrolyzers utilize a proton exchange membrane (PEM), whereas solid oxide electrolyzer cells[9] are steam electrolyzers operating at temperatures between 500 and 850 °C.

In general, for conventional and PEM electrolysis, the respective cell voltages considerably exceed the theoretical decomposition voltage of water electrolysis of 1.23 V at 25 °C and 1 atmosphere.

The plots in Figure 2.4 show indeed that conventional (alkaline) electrolyzers have high overpotential and a relatively small production rate, whereas membrane and advanced alkaline electrolyzers display very similar performance, with lower overpotential and much higher production rates.

Alkaline and PEM electrolysis technologies were the first to be marketed commercially, with the best current electrolyzers requiring an energy input between 4.3 and 4.9 kWh (Nm)$^{-3}$ H_2 (Nm^3 stands for "normal" m^3 under standard pressure and temperature conditions) which corresponds, for the best electrolyzers, to an efficiency of 80–85%.

Alkaline water electrolysis eliminates the need for the expensive precious metal catalysts required for acidic electrolysis and is the technology used in current practice for large-scale electrolytic hydrogen production. Potassium hydroxide enhances the electrical conductivity of water, so that ions can be transported through the electrolyte during electrolysis (Figure 2.5).

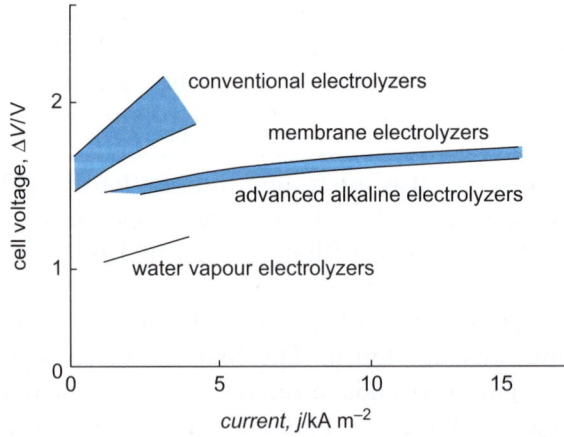

Figure 2.4 Range of performance of different water electrolyzers: Cell voltage divided by current density.
(Image courtesy of Prof. S. Trasatti.)

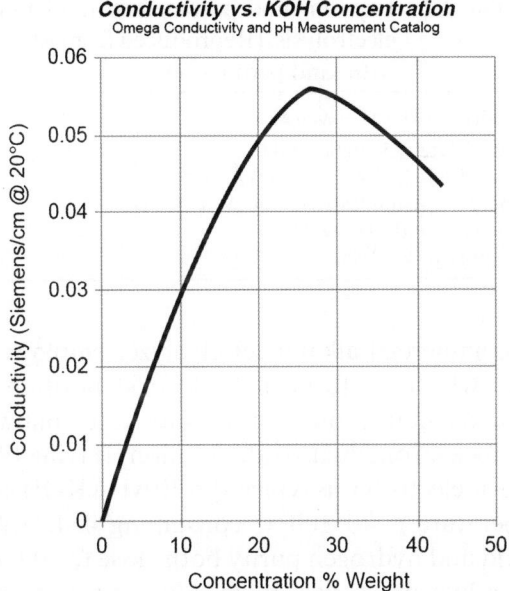

Figure 2.5 Conductivity of the KOH electrolyte as a function of weight percent KOH in water, on right.

In an alkaline electrolyzer, the anodic and cathodic reactions are:

$$\text{Anode: } 2\text{OH}^-(\text{aq}) \rightarrow \tfrac{1}{2}\text{O}_2(\text{g}) + \text{H}_2\text{O}(\text{l}) + 2\text{e}^- \quad E_{25°C} = 1.299 \text{ V} \quad (2.11)$$

$$\text{Cathode: } 2\text{H}_2\text{O}(\text{l}) + 2\text{e}^- \rightarrow \text{H}_2(\text{g}) + 2\text{OH}^-(\text{aq}) \quad E_{25°C} = 0.00 \text{ V} \quad (2.12)$$

Hence, the net reaction is:

$$\text{H}_2\text{O}(\text{l}) \rightarrow \text{H}_2(\text{g}) + \tfrac{1}{2}\text{O}_2(\text{g}) \quad E_{25°C} = 1.299 \text{ V} \quad (2.13)$$

The anodic reaction for O_2 evolution of water electrolyzers is particularly demanding because its mechanism involves three or more steps, and the intermediates are species of high energy, involving high activation energies.[10]

In alkaline solution, the discharging particles that produce H_2 evolution at the cathode are H_2O molecules, and because hydrogen evolution produces hydroxyl ions, a strongly alkaline medium always develops close to the cathode surface.

Table 2.2 Experimental parameters in alkaline electrolysis. (Reproduced from Ref. 3, with kind permission.)

$H_2O(l) \rightarrow H_2(g) + \frac{1}{2}O_2(g)$
Electrolyte: 25–30% KOH
$\Delta V = 1.65–2.00$ V, $j = 1–10$ kA m^{-2}
Energy consumption: 4–4.9 kWh m^{-3} H$_2$
Current yield: 98–99.9%
H$_2$ purity: >99.8%

Typically, a commercial alkaline electrolyzer employs porous Raney nickel electrodes which are formed by electrodeposition of a Ni–Al or Ni–Zn alloy onto a metallic (often mesh) substrate, followed by leaching of the Al or Zn by a strong hydroxide solution, leaving behind a porous Ni structure.[11] The electrolyte is typically a 30 wt% KOH solution and the operating temperature is 70–100 °C consuming 4–4.9 kWh m^{-3} of H$_2$, with current yield and hydrogen purity both close to 100% (Table 2.2).[12]

The potassium hydroxide electrolyte solution is very corrosive to the skin and can cause blindness if it comes in contact with the eyes, in addition to other serious symptoms. For example, members of the HARI[13] project team in the UK who were exposed to KOH fumes for long periods experienced throat irritation and, although they were not listed as symptoms on safety sheets for this chemical, intense headaches and tiredness. The use of specific alkali-resistant breathing masks was, therefore, necessary (Figure 2.6).

The durability of alkaline electrolyzers is sufficiently high, giving a typical operating life of 10–20 years.[14] The lack of a membrane, however, means that hydrogen cannot be produced at high pressure and the highly concentrated KOH electrolyte is sensitive to contamination.

Originally developed as part of the "Gemini" space program by Brown, Boveri Ltd (today's ABB) over the years from 1976 to 1989,[15] the proton exchange membrane electrolyzer technology utilizes a cation exchange electrolyte membrane (normally Nafion from DuPont) between the cathode and anode compartments.

Under these conditions, hydrogen pressurized up to 75 bar can be produced easily, with higher efficiency than alkaline electrolysis. The plots in Figure 2.7, for example, show that at 200 mA cm^{-2}, an experimental PEM electrolyzer has a voltage efficiency of 95%, compared with 87% per cell for a commercial alkaline (Teledyne) electrolyzer, while at a higher current density of 350 mA cm^{-2}, the difference in relative efficiency is even greater (92% for the PEM and 81% for the Altus 20).[16]

However, the highly corrosive environment of PEM electrolyzers necessitates the use of precious metal catalysts such as Pt and Ir for the

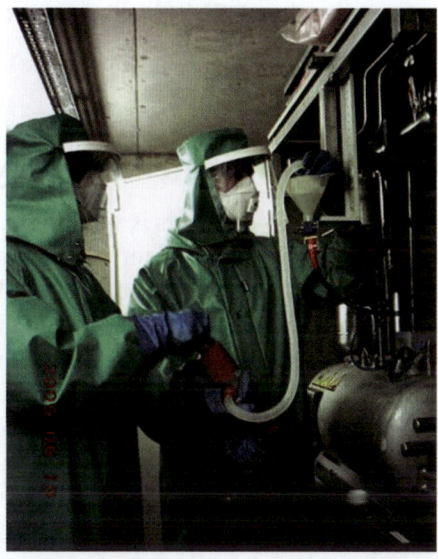

Figure 2.6 Safety equipment being used by members of the HARI renewable hydrogen project team while handling potassium hydroxide electrolyte at West Beacon, UK.
(Reproduced from Ref. 13, with kind permission.)

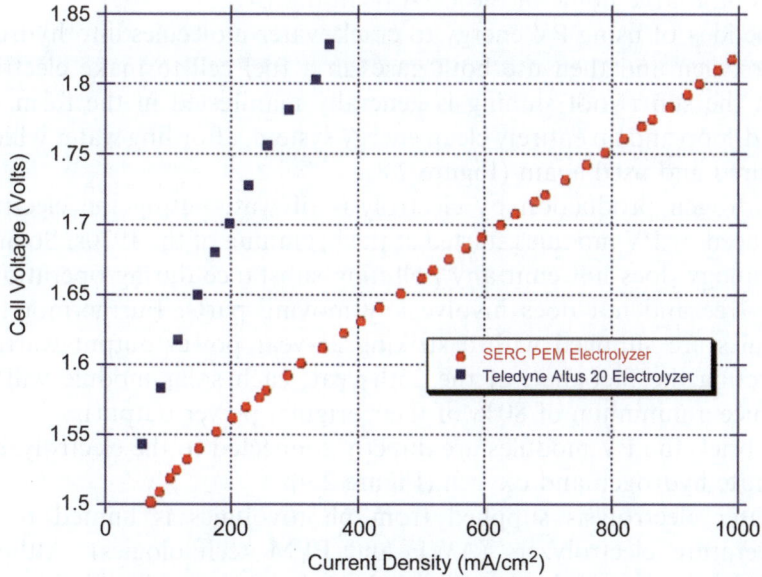

Figure 2.7 Comparative cell voltage–current density graph of a PEM and commercial alkaline electrolyzer at a hydrogen output pressure of 100 psig. This translates into more hydrogen generated per unit cell area, and thus to more compact and economical electrolyzer design.
(Reproduced from Ref. 16, with kind permission.)

cathode, which of course makes PEM electrolyzers considerably more expensive than AWE technology.

2.3 Photovoltaic-Assisted Water Electrolysis

At first sight, given that the cost of water is negligible, the economics of the process of hydrogen generation by water electrolysis is driven by the cost of electricity and by the cost of the electrolyzer. However, because the cost of photovoltaic (PV) electricity consumed has been more valuable for decades than the hydrogen produced, this method has *not* been used. Hydrogen generated by water electrolysis, however, is an ideal way to store intermittent solar electricity generated during the day. The advantages of hydrogen as a storage medium are self-evident: *i*) high specific energy; *ii*) low or zero self-discharge rate (H_2 can be stored for years, unlike other energy storage media); *iii*) it is clean, because no pollution is produced.

Solar hydrogen can therefore be used to fuel the power needs of homes, vehicles or boats, thus enabling decentralized energy generation. For example, 11.4 kg (3 gallons) of water, once split into O_2 and H_2, contains enough energy, when recombined, to satisfy the daily energy needs of a large home in the USA or in the EU.

The idea of using PV energy to crack water molecules into hydrogen and oxygen and then use both gases in a fuel cell to make electricity when the sun is not shining is generally manifested in the form of a closed-loop and an entirely clean energy system, affording water which is captured and used again (Figure 2.8).

Hydrogen production by electrolysis of water using the electricity produced by PV modules started at the beginning of the 1970s. Solar PV technology does not emit any polluting substance during operation, is noise-free and not does involve any moving parts. Furthermore, PV modules are supplied with a striking 25-year power output warranty (reflecting the fact that, at the 25th year, each solar module will still produce a minimum of 80% of their original power output).

In brief, the PV modules are directly connected to the electrolyzer to generate hydrogen and oxygen (Figure 2.9).

Water electrolysis supplied from photovoltaics is limited to low temperature electrolyzers (AWE and PEM technologies). Although AWE is a mature and robust technology, its corrosive liquid electrolyte and less compact designs mean that PEM technology is a more promising WE electrolysis format for direct coupling with renewable electrical sources.[17]

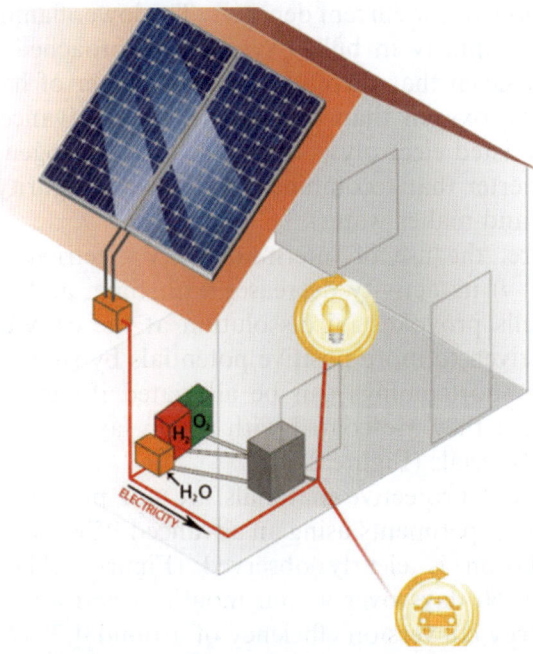

Figure 2.8 During the day, PV modules power the home. At the same time, excess energy is used to split water into H_2 and O_2 for storage and subsequent usage in fuel cells.
(Adapted from MIT, with kind permission.)

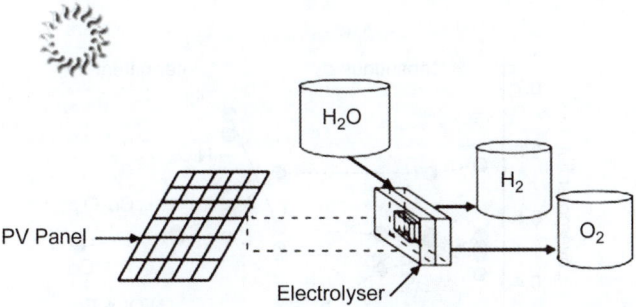

Figure 2.9 Schematic of a PV hydrogen (and oxygen) production system.
(Reproduced from Ref. 17, with kind permission.)

The purity of the O_2 and H_2 gases produced by an alkaline electrolyzer is affected by the current density and temperature of the cells.[18] In detail, the purities of the hydrogen and oxygen gases are poorer at low current densities (such as when a cloud covers the sun): diffusion of the gases through the liquid electrolyte is a more significant fraction of the total production at low current densities. The lower flammable limit, 4% for hydrogen impurity in bulk oxygen, is approached at low current densities, and given that there is a greater danger of having hydrogen impurity in the oxygen than the reverse, most advanced electrolyzers used in PV-assisted electrolysis make use of a hydrogen gas purifier (a catalytic converter that recombines any oxygen impurity in the hydrogen product, and makes water).

Furthermore, the use of intermittent PV electricity results in two shortcomings: *i*) its activity decreases with time, and *ii*) shutdown of electrolytic cells provokes Ni dissolution at the cathode because this electrode is driven to more positive potentials by short-circuit with the anode. These shortcomings can be alleviated if the Ni cathodes are *activated*, *i.e.*, if they are coated with a thin layer of more active and more stable materials (Figure 2.10).

When the said protective materials are not present, even in recent direct coupling experiments using an advanced PEM electrolysis system, stack degradation is clearly observed (Figure 2.11), affording for approximately 60 days over a four-month period an overall solar-to-hydrogen energy conversion efficiency of around 4.7%.[19]

The connection between the solar generator and the electrolyzer can either be direct, by feeding the electrolyzer with direct current (DC) generated by the modules,[20] or, more efficiently, can be mediated by an electronic, instantaneous match between the maximum power point (MPP) of the solar generator and that of the electrolyzer.

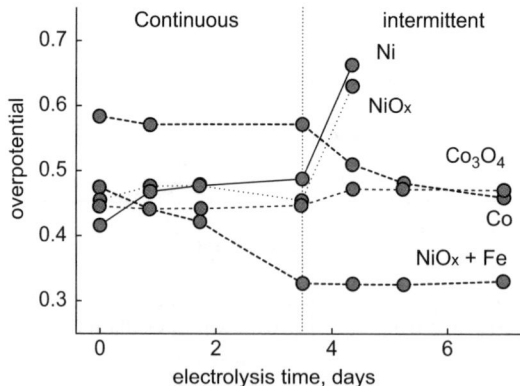

Figure 2.10   Variation of overpotential (inversely proportional to activity) for O_2 evolution as a function of time for continuous and intermittent electrolysis. Under the latter conditions, Ni-based cathodes need to be protected by a thin layer of more active and more stable materials. (Image courtesy of Prof. S. Trasatti.)

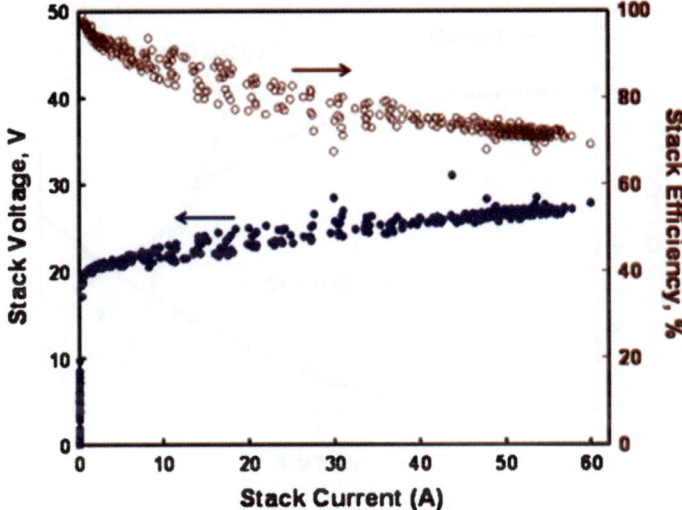

Figure 2.11 Voltage–current data and electrolyzer efficiency collected in real time over a period of five days' operation for a 13-cell PEM electrolyzer stack in the very early stages of the direct coupling experiment with a 2.4 kW PV array. (Reproduced from Ref. 19, with kind permission.)

In other words, to maximize hydrogen generation in a PV-electrolysis system the operating point of the total system must equal the MPP of the solar generator. This is usually realized by an MPP tracker that guarantees the operation of the solar generator at its maximum power point. In addition, a DC/DC converter shifts the power to the characteristic of the load, adapting the output of the solar generator to the input of the electrolyzer (P_3 in Figure 2.12, where the labeled PV system and electrolyzer curves represent the I, V characteristics of the PV and electrolyzer systems, respectively).

Figure 2.13 shows the hydrogen flow rate measured for a sample day in July in the case of coupling with the MPPT and, for comparison, the hydrogen flow rate in the case of direct coupling and coupling with the MPPT in relation to the variation of solar radiation intensity. Clearly, the MPPT optimizes the system performance, increasing the system current for the same radiation intensity, which leads to greater hydrogen production.

However, Paul and Andrews have recently demonstrated the possibility of achieving near maximum power transfer between a directly coupled PV array and a proton exchange membrane (PEM) electrolyzer stack by finding an optimal configuration of the series–parallel connection of both the PV modules and the PEM cells.[21] This entirely

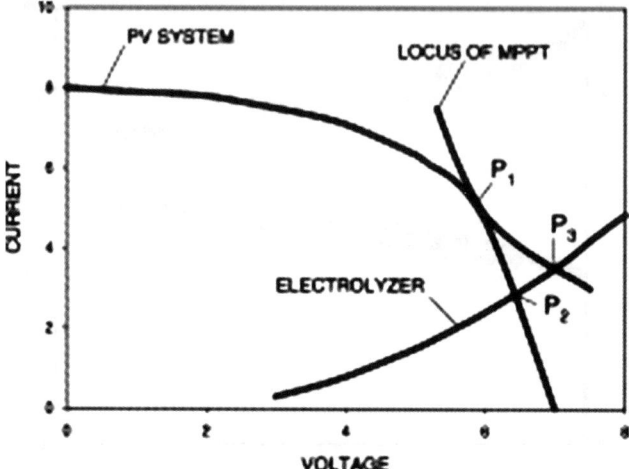

Figure 2.12 Characteristics of PV and electrolyzer systems. The line labeled "locus of
MPPT" shows the maximum power for a given radiation intensity.
(Reproduced from Ref. 19, with kind permission.)

avoids the costs of an electronic coupling system (around $ 500 per kW,
as of the end of 2011).

Table 2.3 shows that the application of this procedure to four (75 W) PV
modules directly coupled to five 50 W PEM electrolyzer stacks (250 W in
total) predicts an energy transfer of over 94% of the theoretical maximum,
while experiments indicate an actual energy transfer of around 95%.

The approach has been generalized recently to afford a new method
for relative sizing of both components, based on simple modeling of
both polarization (for the electrolyzer and the PV array) curves.[22]
Modeling and simulation is used to extract a cloud of maximum power
points under all the radiation and temperature conditions for a normalized PV generator. Subsequently, the ideal ratio between the sizes of
the components is obtained by fitting a normalized polarization curve
for the electrolyzer to this cloud of maximum power points. Direct
coupling of an advanced electrolyzer to a matched solar PV source for
hydrogen generation and storage, involving minimum interfacing electronics, does lead to substantial cost reduction and enhances the economic viability of solar-hydrogen systems.

In general, the overall efficiency of the system (η_s) is given by Equation
2.14:[23]

$$\eta = \frac{QE}{HA_m} \tag{2.14}$$

Where Q is the hydrogen flow rate (mL s^{-1}), E is the calorific value for
hydrogen (as a net or gross calorific value, 10.8 J mL^{-1}, and 12.7 J mL^{-1}

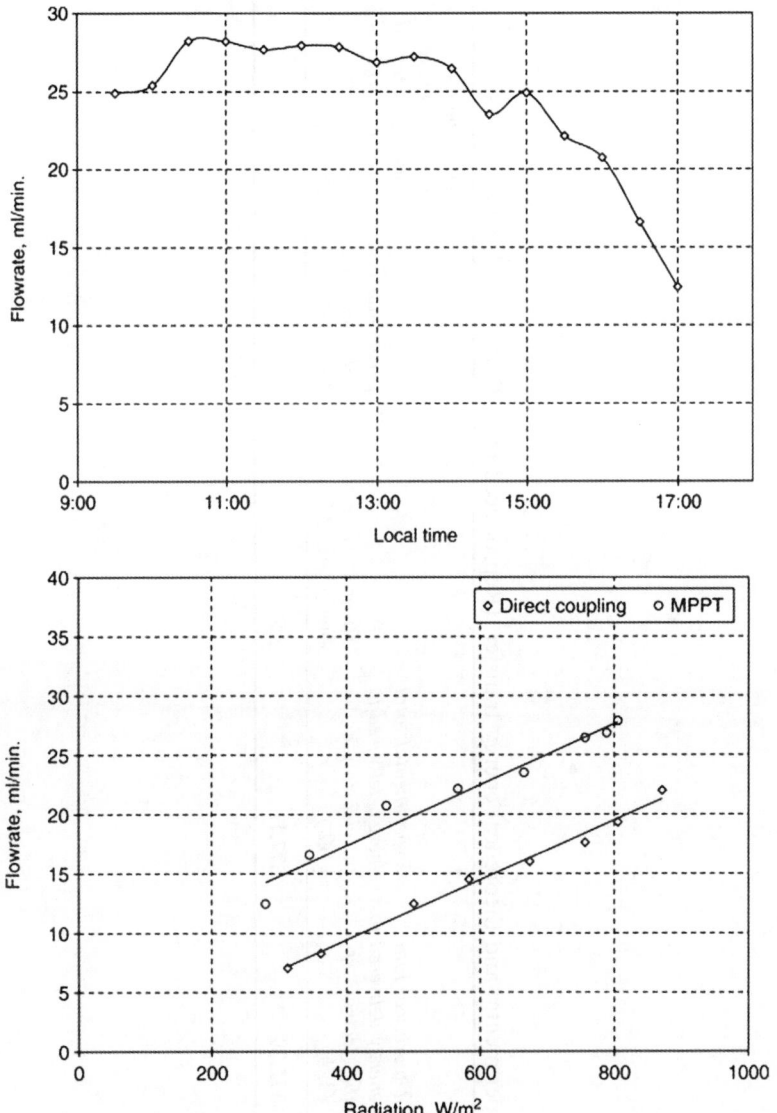

Figure 2.13   Hydrogen flow rate measured for the system with MPPT (top). Hydrogen flow rate versus solar radiation measured for the direct coupling system and for MPPT (lower figure).
(Reproduced from Ref. 19, with kind permission.)

at 0 °C and 1 atm), H is the solar radiation intensity (W m^{-2}) and A_m is the photovoltaic module area (m^2). In other words, the efficiency is defined as the ratio of the higher heating value of hydrogen produced in one year to the yearly total solar energy on the PV modules.

Table 2.3 Comparison of experimental and theoretical energy transfer. (Reproduced from Ref. 21, with kind permission.)

Total direct coupling time (h:min:s)	Maximum total PV energy available (Wh)	Theoretical total energy delivered to the electrolyzer (Wh)	Experimental total energy delivered to the electrolyzer (Wh)	Theoretical overall energy loss $\Delta E\%$	Experimental overall energy loss $\Delta E\%$	Discrepancy between theoretical and experimental energy transfer (% total energy delivered)
2:48:20	395.30	372.29	377.11	5.82	4.60	1.28

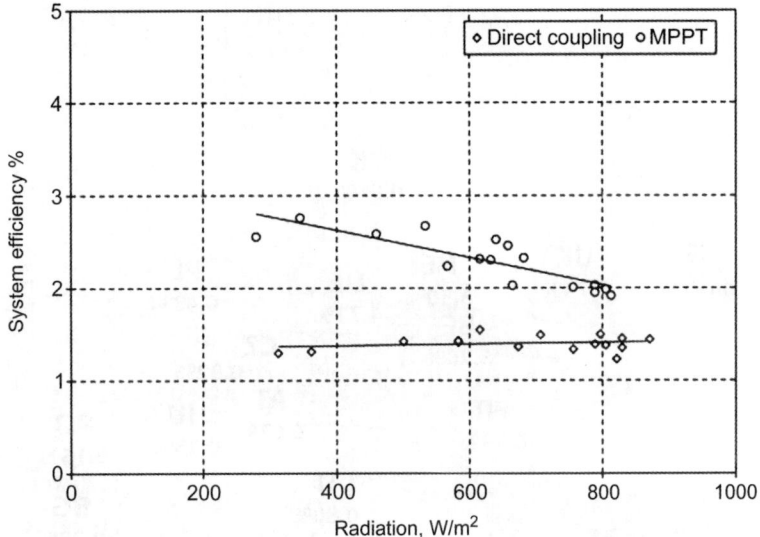

Figure 2.14   Overall system efficiency versus solar radiation measured for the direct coupling system and for MPPT.
(Reproduced from Ref. 23, with kind permission.)

The overall typical system efficiencies in the case of direct coupling and with the MPPT are shown in Figure 2.14. Although the *electrolyzer* efficiency decreases slightly with the MPPT, due to the higher electrolyte temperature and higher ohmic losses, the overall *system* efficiency increases in comparison with the direct coupling case. The system operates around the maximum power point of the PV module, and the maximum power point tracker increases the system current and accordingly increases the hydrogen flow rate.

In general, with the efficiency of modern photoconverters and electrolyzers being about 20% and 80%, respectively, the total efficiency of solar radiant energy transformed to chemical hydrogen energy is nearly 16%.[24] Actually, the overall efficiency is of the order of 10%.[25] Indeed, Gibson and Kelly demonstrated a total PV-H$_2$ system efficiency of 12.4% by optimizing the choice of the PV module directly coupled to a PEM electrolyzer working at around 31.7 V and 4.7 A at nominal conditions.[26]

Now, some argue that PV electricity, with its typical 200 W solar modules measuring 1.2 m^2, would require immense consumption of land to cover even a fraction of the power demand of economically advanced countries.

This is simply not the case. On average, for example, covering 0.6% of the European territory by obsolete, 10% efficient PV modules would

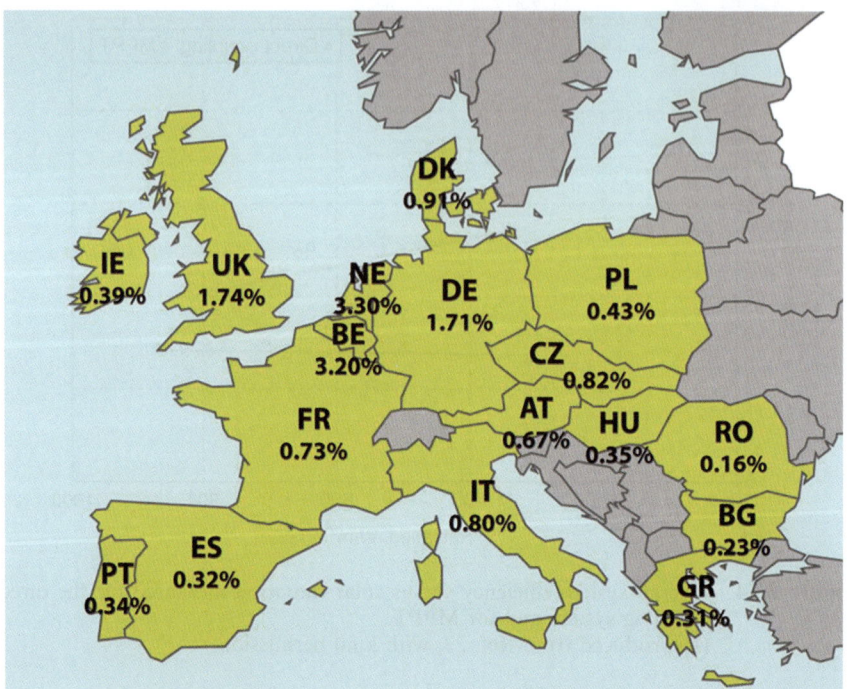

Figure 2.15   Theoretical PV potential: surface of PV modules (10% conversion efficiency) mounted at the optimum angle that would be needed to completely
satisfy the electricity consumption of some selected European countries,
expressed as % of the country's area. The European average is 0.6%.
(Reproduced from Ref. 27, with kind permission.)

theoretically satisfy its entire electricity demand (Figure 2.15).[28] This
2007 estimate was largely conservative because the latest PV module
technologies have about 18% conversion efficiency, greater than that
considered in the cited analysis (10%), so the area covered per kWp is
roughly half and will certainly continue to decrease in the future. Furthermore, the PV area does not translate directly into land area covered
because a large share of PV panels (currently about 90%) are and will
normally be placed on rooftops, which would exist regardless of whether
solar panels are installed.

2.4   Economics of Water Electrolysis

A detailed cost analysis, performed in 2004,[29] of the domestic production of hydrogen using a photovoltaic-electrolyzer system showed that,
for a 1 kWp photovoltaic system with fixed modules, depending on the

annual solar radiation on a horizontal surface H_T, the cost of hydrogen varies from 3.5 to 38 $\$\,\text{kg}^{-1}$ with a corresponding energy cost from 26 to 268 $\$\,\text{GJ}^{-1}$. Specifically, the hydrogen energy costs C_{HM} (in $\$\,\text{kg}^{-1}$ H_2) and C_{HE} (in $\$\,\text{GJ}^{-1}$ H_2) are correlated empirically with the price of the PV plant *and* of the electrolyzer, expressed in $\$$ per W_p, (P_{PV} and P_{EL}, respectively) according to the following equations:

$$C_{HM} = 12.34[(P_{PV} + P_{EL})/H_T]^{0.85}$$
$$0.25 < (P_{PV} + P_{EL})/H_T < 4 \tag{2.15}$$

$$C_{HE} = 87.79[(P_{PV} + P_{EL})/H_T]^{0.85}$$
$$0.25 < (P_{PV} + P_{EL})/H_T < 4 \tag{2.16}$$

In 2004, the price of energy for gasoline engine powered vehicles was about $(0.5\,\$\,\text{L}^{-1})/(0.73\,\text{kg}\,\text{L}^{-1} \times 0.046\,\text{GJ}\,\text{kg}^{-1}) \approx 15\,\$\,\text{GJ}^{-1}$. Hence, even considering that a fuel cell powered vehicle is more efficient than a gasoline engine powered vehicle, Bilgen concluded that, for solar hydrogen:

> The lower end price of about 25 $/GJ may just be competitive in heavily subsidized situations. Otherwise it is clear that as expected, with the present photovoltaic and electrolyzer price structure, domestic produced hydrogen will *not* be competitive with the fossil fuel derived gasoline or similar fuels.

Seven years later, however, the price of gasoline in Northern America in 2010 was $0.76\,\$\,\text{L}^{-1}$,[30] *i.e.* $22.63\ \$\,\text{GJ}^{-1}$, whereas the photovoltaic-electrolyzer system price is about *10 times* lower than in 2004.

Using the Hydrogen Analysis (H2A)[31] tool, a standard methodology developed by the US Department of Energy (DOE) to estimate hydrogen production costs, in 2006, scientists at the US National Renewable Energy Laboratory reached the conclusion that in order to meet the DOE's cost target for hydrogen at $2.00\text{–}3.00\ \$\,\text{kg}^{-1}$, electrolyzers with 2006 efficiencies would need to have access to electricity prices lower than $0.045\text{–}0.055\ \$\,\text{kWh}^{-1}$ (Figure 2.16).[32]

According to the state of the art technology of 2006 electrolyzers, ideal systems would need to have access to electricity prices lower than $\$0.075\,\text{kWh}^{-1}$, representing the highest possible electricity price in 2006 that low-temperature electrolyzers would be able to use to produce hydrogen at $\$3.00\,\text{kg}^{-1}$.

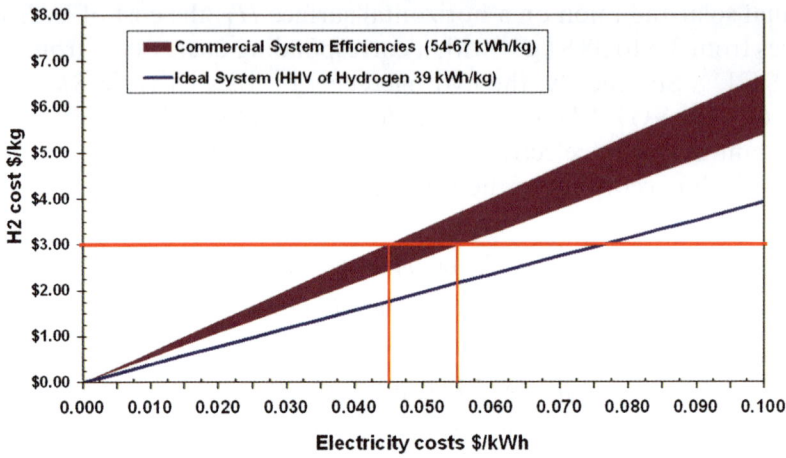

Figure 2.16   Cost of hydrogen versus electricity without equipment costs.
(Reproduced from Ref. 32, with kind permission.)

2.5   Reducing the Cost of Electrolytic Hydrogen

In a thoughtful and somehow prophetic analysis[33] of photovoltaic-assisted water electrolysis dating back to 1982, Carpetis concluded that:

> There are three interconnected subsystems (solar array, electrolysis unit and hydrogen storage unit) which should be optimized for minimum hydrogen costs, according to the local conditions and to the hydrogen utilization schedule...

> The cost optimization and the break-even conditions will depend not only on the solar array and electrolyzer performance improvement, but also on the location of the production and the hydrogen utilization schedule.

Observing that, at that time, the electrolysis unit costs contributed "a relatively low part" of the total costs, he continued that, in the near future, the cost reduction for hydrogen produced by solar electrolysis had to be expected as a consequence of the PV array cost reduction (Figure 2.17).

However, in the long term, further cost reduction could be expected owing to the use of more efficient solar cells and electrolytic units. With lower production costs, low cost storage methods for solar hydrogen will become more important.

The forecast of Carpetis turned out to be correct. Following a true collapse in the cost of solar electricity, a rapid and concomitant decrease

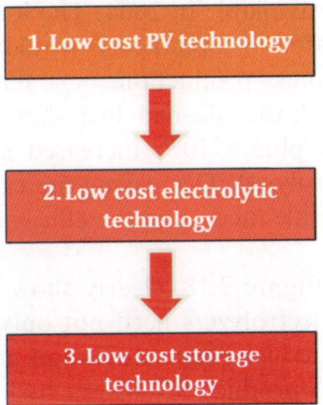

Figure 2.17 Consecutive steps in the cost reduction trend of hydrogen produced by solar water electrolysis.
(Created according to Ref. 28.)

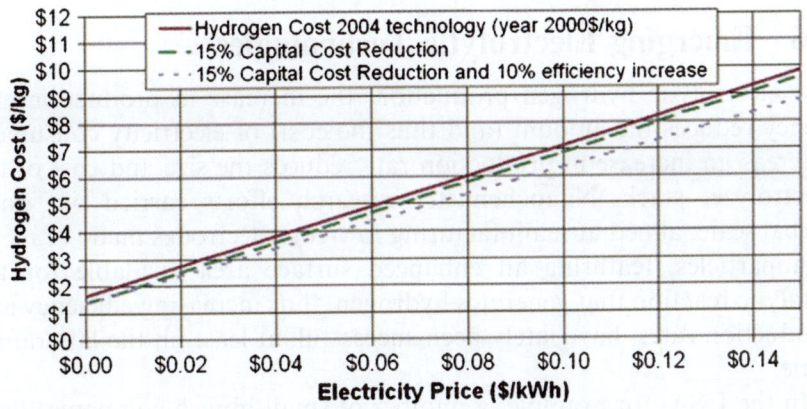

Figure 2.18 How hydrogen cost varies with electricity price. The linear relationship between electricity price in $\$\,kWh^{-1}$ and hydrogen cost in $\$\,kg^{-1}$. (Reproduced from Ref. 34, with kind permission.)

in cost and technology improvement are currently affecting the electrolyzer. Significant improvements have been made, making it possible to reach improved cell efficiencies and higher current densities at far lower cost compared with previous technologies.

The graph in Figure 2.18 displays three scenarios of how the calculated price of hydrogen varies according to electricity price,[34] as well as with the cost of electricity (including capital, operating and maintenance costs) for electricity prices up to $0.15 kWh^{-1}. The solid line displays how the cost of hydrogen changes with electricity prices using technology and prices available in 2004. The longer-dashed line shows the effect a 15% reduction in capital costs would have on hydrogen cost (as a consequence of mass production, or a simplification of the auxiliaries); whereas the shorter-dashed line shows the effect of a 15% capital cost reduction plus a 10% increased system efficiency (from improvements in electrolysis). Decreasing the capital costs changes the intercept of the line, while increasing the efficiency changes the *slope* of the line.

The same plots, in Figure 2.18, clearly show that to be competitive with gasoline prices, electrolyzers need not only to obtain inexpensive electricity, but also to reduce *capital* costs and improve the *efficiency* of the systems. Indeed, to produce $2.00 kg^{-1} hydrogen, electricity prices will need to be available for 0.007, 0.011 and 0.012 $ kWh^{-1} with, respectively, 2004 technologies, a 15% capital cost reduction, and a 15% capital cost reduction *plus* a 10% improvement in efficiency.

2.6 Emerging Electrolytic Technologies

For electrolytic hydrogen production, the increase in production efficiency reduces the amount (and thus the cost) of electricity consumed, whereas an increase in production rate reduces the size and cost of the electrolyzer stack. Nanochemistry research efforts carried out on a global scale, aimed at manufacturing low-cost electrodes made of metal nanoparticles, featuring an enhanced surface area available for the catalytic reaction that generates hydrogen, thus increasing efficiency and production rates, have lately been successful, at least on the laboratory scale.

In the USA, for example, a number of small hi-tech companies have developed new Ni-based catalysts. *QuantumSphere* manufactures a *Nano NiFe* coating of nickel and iron nanoparticles capable of increasing the efficiency of an alkaline electrolyzer using coated cathodes by more than 10%.[35] The company's experimental electrolyzer stack produces 2.8 Nm^3 of hydrogen per day at 68% efficiency in normal operation, close to the 2012 target of 69% efficiency for advanced electrolyzers set by the Department of Energy.

Another small company, *GridShift, Inc.*, uses a coating technique that coats all surfaces of a three-dimensional shape (like reticulate nickel foam) with a nano catalyst exposed to the electrolyte's boundary layer, to generate hydrogen in an electrolyzer running at 80% energy efficiency with a current density of 1000 mA cm^{-2} overall, delivering compressed hydrogen at around $2.51 kg^{-1} H$_2$.[36]

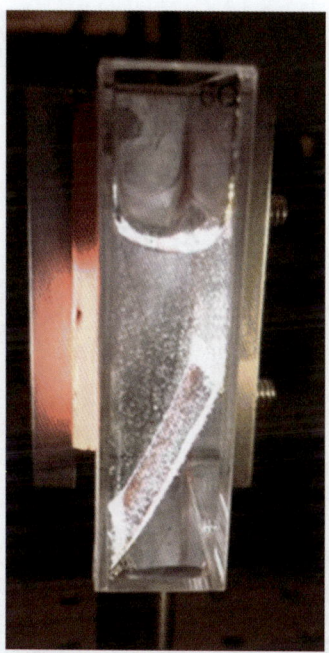

Figure 2.19 The cobalt oxygen-evolving catalyst deposited on the ITO-passivated *p*-side of an *np*-silicon junction enables the majority of the voltage generated by the solar cell to be utilized for driving the water-splitting. In the laboratory, the system worked continuously for three days. (Reproduced from Ref. 41, with kind permission.)

Similarly, *Sun Catalytix*[37] is currently trying to commercialize a nickel borate catalyst[38] for the oxygen evolution reaction. The cobalt oxygen-evolving catalyst[39] (Figure 2.19) originally reported gave a current density of $1\,mA\,cm^{-2}$ with an overpotential of $410\,mV$, which is worse than the typical performance (*e.g.* $1\,mA\,cm^{-2}$ at $<200\,mV$ overpotential) for nickel anodes.[40]

These and other companies are seeking financial support to scale up their processes. In any case, the trend towards reducing the cost by enhancing the efficiency of electrolysis is clear, and a number of new companies are emerging that have commercialized ever more efficient electrolyzers.

New technology recently developed in Italy combines the advantages of the cheap, nickel-based electrode materials used in alkaline electrolysis with the production of hydrogen at the high pressures typical of solid polymer membrane PEM electrolyzer technology (Figure 2.20). With no liquid electrolyte on the H_2 side, hydrogen is produced at higher purity.

The resulting home generator produces hydrogen safely on demand from water, directly compressed, dry and pure (Table 2.4), providing the

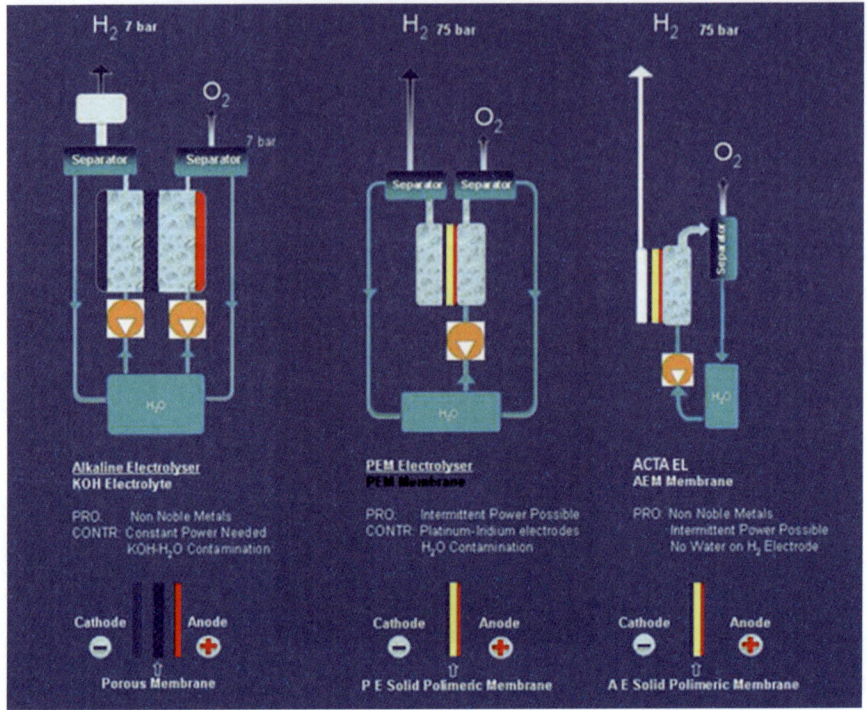

Figure 2.20 Scheme of alkaline (*left*), PEM (*middle*) and innovative PEM (*right*)
electrolysis.
(Reproduced from Actaenergy.it, with kind permission.)

Table 2.4 Requirements of the H_2 produced by Acta's
EL100 electrolyzer. (Reproduced from
Actaenerrgy.it, with kind permission.)

Production	$100 \, L \, h^{-1}$
Pressure, bar	15 (30)
Purity	99.95%
Power consumption	550 W
Water consumption	$0.085 \, L \, h^{-1}$
Width	25 cm
Height	46 cm
Depth	50 cm

ideal refill for fuel cell applications on the market that require compressed hydrogen for reasons of energy density.

At the end of 2011, the company shipped its first hydrogen generator
stack that can produce 500 L of hydrogen per hour to an Italian engineering firm that specializes in industrial heating systems.[42]

Figure 2.21 ITM Power's stack, featuring proprietary membrane materials, for the *HFuel* generator.
(Reproduced from hfcletter.com, with kind permission.)

Similarly, in the UK, the company ITM Power has developed an innovative alkaline solid polymer membrane platform for its electrolyzer (Figure 2.21), which achieved 57% cost savings compared with their PEM-based stack, due to the removal of Pt catalysts and the simplification of the system.[43]

The company commercializes an on-site *HFuel* generator for refueling hydrogen-powered vehicles that produces hydrogen by electrolysis, compresses it, stores it and dispenses the gas on demand at high pressure (350 bar). The new ITM electrolyzer membrane platform transports OH^- rather than H^+ ions, enabling smaller stack sizes as a result of the higher ionic conductivity that makes high current densities achievable, while the high water permeability allows considerable simplification of the water management system.

Also in the UK, RE Hydrogen has developed a low cost, ambient pressure alkaline electrolyzer (Figure 2.22) based on a 5 kW stack module, using plastic materials and a proprietary regenerative carbon aerogels-based catalyst, whose retail price is estimated to be 70% lower than the current market price for conventional alkaline electrolyzers.[44] In detail, the low cost cathode consists of resorcinol–formaldehyde (RF) carbon aerogels of high surface area ($> 700 \, m^2 \, g^{-1}$) and nano-pore sizes (4 nm) thermally deposited on molybdenum metal.[45]

The scanning electron microscope (SEM) images of the Mo–RF electrode (Figure 2.23), where the hydrogen evolution reaction takes place at 298 K, indicate formation of a highly porous carbon

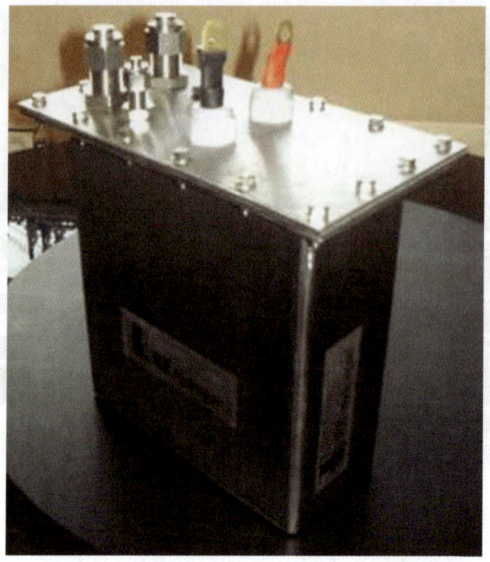

Figure 2.22 The new electrolyzer developed by RE Hydrogen in the UK.
(Reproduced from rehydrogen.com, with kind permission.)

RF coating on Mo-metal mesh.   Fine pores of RF carbon

Figure 2.23 SEM image of coated Mo–RF electrode indicating effectiveness of
synthesis method.
(Reproduced from Ref. 43, with kind permission.)

nanostructure whose effectiveness in alkaline water electrolysis is clearly
shown by the data in Table 2.5, wherein the charge-transfer resistance of
the Mo–RF electrode is reduced by about 70% when compared with
pure molybdenum metal.

Table 2.5 Solution resistance, R_s, charge-transfer resistance, R_{ct}, and double-layer capacity, C_{dl}, derived from analysis of impedance spectra recorded at $E = 1.5\,V$ in 30% by vol. KOH solution at 298 K. (Reproduced from Ref. 43, with kind permission.)

Electrode	Voltage (V vs. SHE)	R_s ($\Omega\,cm^{-2}$)	R_{ct} ($\Omega\,cm^{-2}$)	R_{dl} ($\Omega\,cm^{-2}$)
Mo	1.5	0.36	59.96	0.51
Mo-RF	1.5	4.57	17.98	0.37

Table 2.6 Cost of hydrogen using RE Hydrogen electrolyzer subject to different electricity costs.

Electricity price	RE Hydrogen cost of H_2
Green grid electricity at £0.12 kWh^{-1}	£7.7 kg^{-1} green hydrogen Oxygen produced free
Commercial grid electricity at £0.077 kWh^{-1}	£5.5 kg^{-1} hydrogen
Direct solar PV electricity at £0.04 kWh^{-1}	£3.6 kg^{-1} hydrogen High gas purity

Molybdenum is about 3 to 6 times more expensive than nickel but far less expensive than platinum, offering a clear economic benefit in reduced capital cost investment compared with other electrodes (such as Pt–C) previously used in electrolyzers.

Unlike conventional electrolyzers, the RE Hydrogen electrode can now operate under variable and intermittent mode for unlimited on–off switching cycles, owing to its *in-situ* regenerative capability of the electrode-catalyst, which bestows a long life. As a result, the company claims that its electrolyzer can produce "green" hydrogen at a price in the range of £3–7.7 kg^{-1} (subject to the electricity cost), far lower than the current cost of commercial hydrogen, stored in cylinders, which is sold at £12–110 kg^{-1} (Table 2.6).

Thanks to the innovations mentioned above, the company claims to have reduced the capital cost by a remarkable 90%. Under these conditions, then, the cost of hydrogen is mainly influenced by the price of electricity.

The electricity contributes 60% of the cost of hydrogen for RE Hydrogen's system, while for conventional pressurized electrolyzers only 22% of the cost of hydrogen can be ascribed to the cost of the electricity (Figure 2.24). Therefore cheaper or free electricity will bring a greater reduction in hydrogen price via these innovative electrolyzers.

The company's current offering includes 5 kW, 25 kW and 100 kW systems, while larger electrolyzers can be made available on request by adding modular 5 kW electrolyzer stacks.

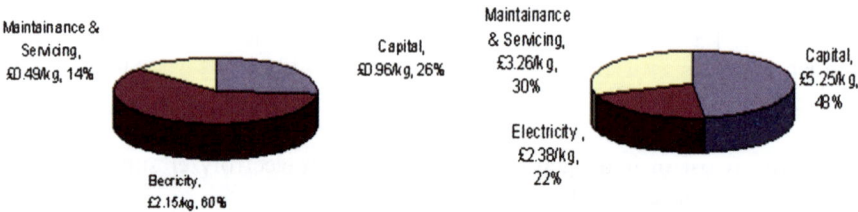

Figure 2.24   Hydrogen cost breakdown for RE Hydrogen (left) and conventional
electrolyzer technologies.
(Reproduced from Rehydrogen.com, with kind permission.)

There are very few manufacturers of hydrogen compressors in the
world, and the cost is often almost *higher* than that of an electrolyzer.
Hence, to increase the marketability of its atmospheric electrolyzers, RE
Hydrogen is currently working to develop a low cost hydrogen compressor that will also be used as a hydrogen dispenser for refueling
vehicles with H_2 compressed at 350 and 700 bar. Thus far, the company
has built a working prototype (Figure 2.25) suitable for a 30 kW electrolyzer for $6 \, Nm^3 \, h^{-1}$ gas flow rate, which is claimed to cost up to 70%
less and to be 50% more efficient than a conventional compressor.[46]

Aiming to establish the UK's first complete solar hydrogen energy
supply chain, RE Hydrogen partnered with fuel-cell manufacturer
Arcola Energy and with Linde's BOC in a collaborative project known
as rabh2.[47] Using RE Hydrogen's 5 kW electrolyzer to reduce the capital
cost of the stack by 90%, while retaining its capability for unlimited on–
off switching cycles (vital for intermittent and variable operation of
renewable energy power), the team will be able to afford to power the
electrolyzer with solar and wind electricity. The high purity hydrogen
thereby obtained will be used to power a wide range of fuel cells built by
Arcola Energy at their factory in East London.

2.7   A Flexible Technology with Large Applicative Potential

Demonstration units such as the Schatz Solar Hydrogen Project stand-
alone energy system, which has powered since 1991 the 600 W air
compressor that aerates the aquaria at Humboldt State University's
Telonicher Marine Laboratory in Trinidad, California, clearly show that

Figure 2.25 The low cost hydrogen compressor developed by RE Hydrogen in the UK. (Reproduced from Rehydrogen.com, with kind permission.)

hydrogen can be used efficiently to store solar energy and that the electrolyzer is flexible enough to respond to the fluctuating solar energy yield with respect to both time and capacity.[48]

The system – the first solar hydrogen energy plant in the USA – consists of a 7.5 kW PV array, a 6 kW alkaline electrolyzer, a 1 kW 120 V AC inverter, and a 1 kW PEM hydrogen fuel cell. During the day, the system uses energy from the sun to power the compressor directly and to produce hydrogen that powers the compressor at night, when the sun is not available (Figure 2.26).

The electrolyzer incorporated into the system was a medium pressure alkaline electrolyzer able to deliver 20 standard liters per minute of hydrogen gas at a current of 240 A at 240 V. The hydrogen gas produced (at a pressure of 7.9 bar) was stored in three conventional tanks with a total capacity of 5.7 m^3 and provided approximately 133 kWh, which operated the load (600 W) for 110 hours, assuming a fuel cell efficiency of 50%. Over eight years of operation each system component had the following efficiencies:[8]

- Faraday 96.4%
- Electrolyzer 79.2%

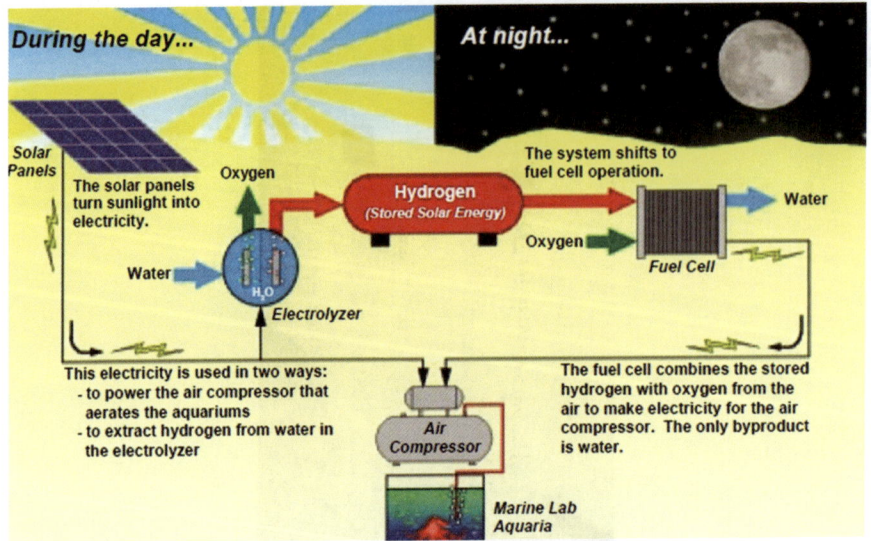

Figure 2.26 Since 1991, a compressor at the Telonicher Marine Laboratory in Trinidad, California has used energy electrolytic solar hydrogen to power the compressor when the sun is not available.
(Reproduced from Ref. 48, with kind permission.)

- Voltage 84.0%
- Fuel cell 43.1%
- Overall electrical storage 34.0%

Further, and much larger, demonstration units such as the German–Saudi *HYsolar* installation, an experimental solar-powered water electrolysis plant operated near Riyadh in the 1990s in a cooperative relationship between Saudi Arabia and Germany, confirmed that the electrolyzer coupled to a PV array is a suitable technology to produce large amounts of clean hydrogen from water and sunlight only.[49]

In the 1990s, however, the retail price of PV modules exceeded 8–10 \$ W^{-1} and each module was losing >1% of the originally rated power. Indeed, after 15 years of operation, in 2006 the Schatz Lab solar hydrogen project was still running, but the PV array that once produced 7.5 kWp was degraded by 16%.[50]

Today modules with linear loss in output power are easily available (Figure 2.27), and the cost of PV electricity has dropped to less than 0.7 \$ W^{-1}.[51] This means that solar electricity has already achieved grid parity with conventional electricity in many regions of the world, including southern Italy, Greece and Spain. It is perhaps not surprising,

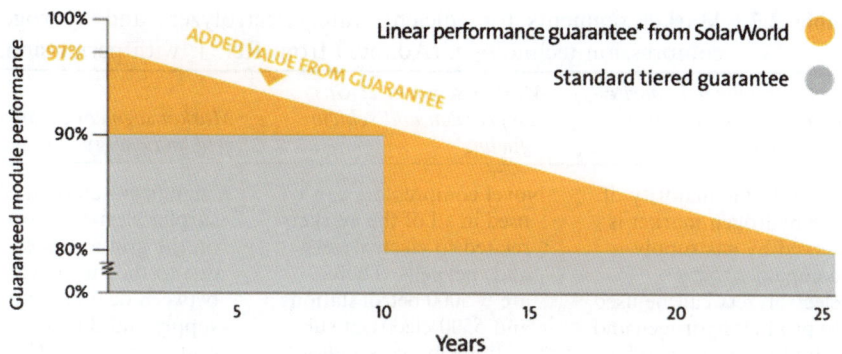

Clear added value compared to standard tiered guarantees.

Figure 2.27 According to this manufacturer's guarantee, the actual power of a new module cannot deviate from the specified rated power by more than 3% during the first year; and afterward, the power will not decrease by more than 0.7% of the rated power per year. So, at year 20, the module's capacity is guaranteed still to be at least 83% of the nameplate. (Reproduced from Solarworld.de, with kind permission.)

Figure 2.28 Low cost generation of hydrogen and oxygen from wind and solar electricity opens the route to multibillion world markets. (Adapted from a figure by Dr A. Roy, with kind permission.)

then, that numerous new commercial solutions to generate hydrogen from PV electricity are eventually reaching the market.

In other words, with low cost electrolyzers and compressors and given the now low cost of PV electricity, described in detail above, *small-scale* hydrogen production (Figure 2.28) becomes a convenient way to capture and store energy from renewable sources and provide fuel for hydrogen-powered vehicles in place of large-scale infrastructure, as well as for

Table 2.7   Market segments for efficient water electrolyzers and hydrogen compression technology. (Adapted from Ref. 1, with permission.)

Market segment 1: onsite hydrogen and oxygen production	Market segment 2: Gas compression and vehicle refueling	Market segment 3: smart grid and energy storage
Currently the majority of the hydrogen market is served by gas supply companies. New electrolyzers can be used to produce hydrogen and refill hydrogen cylinders, and create an end-to-end supply chain network for hydrogen production, distribution and utilization. In the UK alone the estimated size of this market segment is £500 million per year. A relatively large and separate market also exists for oxygen.	Novel compressors can be used in all of the markets related to electrolyzers and fuel cells. There are > 5000 petrol stations and 5500 electrical sub-stations in every major EU country, each of which could potentially install hydrogen compressors for energy storage applications. The fast expanding fuel cell market will need a green hydrogen supply and compressed gas storage.	A significant amount of surplus electricity exists on the grid at times, due to the mismatch between demand and supply and due to the grid constraints. The surplus of wind and PV that cannot be dispatched from solar and wind farms due to the grid's inability to absorb power at a given time can be made available to electrolyzers for storage, and later on used to supply 24/7 electricity.

industrial usage, opening the route to multibillion world markets. New, efficient water electrolyzers and hydrogen compression technology are suitable for several market segments, from energy storage, grid balancing, solar fuel-based power generation, and transport applications (Table 2.7).

There is little doubt that hydrogen produced by solar and wind farms will help to balance the wholesale electricity price, providing also a crucial service for grid stability.

For example, in Canada, a manufacturer of hydrogen generation and fuel cell products, *Hydrogenics*, recently completed a successful trial with an Ontario electricity utility, demonstrating the viability of electrolyzer technology for utility-scale grid stabilization (Figure 2.29).[52] During the trial period, the load from the company's electrolyzer provided frequency regulation in Ontario by responding to power regulation signals from the utility on a second-by-second basis, thus demonstrating *how* hydrogen electricity ensures better balancing of electrical supply and demand while alleviating local transmission constraints. The company announced that it will apply lessons learned from the Ontario project in the development of megawatt-scale energy storage applications.

Figure 2.29 The HySTAT electrolyzer from Hydrogenics was used in Ontario to demonstrate that hydrogen energy is an excellent new way to balance supply and demand to and from the grid.

Another interesting commercial product that uses electrolytic solar hydrogen as fuel is the *Riviera 600* electric boat.[53] With a range of 80 km with a full hydrogen tank, the boat is 6 m long, 2.2 m wide and weighs 1400 kg. The 47% efficiency of the noise-free 4 kW fuel cell engine should be compared to the 18–20% efficiency of a conventional internal combustion engine.

Refueling with compressed hydrogen is relatively fast and simple (Figure 2.30). The boat's fuel system consists of a 20 kg cartridge that can be charged with up to 0.7 kg of hydrogen kept at 350 bar. Refueling is done using a standard filler, plus simple exchange of an empty cartridge for a full one. Compared with battery-powered electric boats, the hydrogen-powered electric boat requires only five minutes to refuel, whereas for conventional electric boats 6–8 hours are typically required to recharge the spent batteries. Moreover, the *Riviera 600* electric motor has twice the range of conventional battery-powered boats.

To generate electricity, a hydrogen refueling station makes use of PV modules integrated in a 250 m² flat roof (Figure 2.31), further connected to an electrolytic cell. Even at Austria's cold latitudes this station is capable of producing an annual yield of 823 kg hydrogen, equivalent to 1100 cartridges with a 27 200 kWh energy content, enough hydrogen to run a boat for 80 000 km. The "Future Project Hydrogen" team has created budget calculations for the generation of hydrogen on-site by use of PV cells on the premises of 10 boats for commercial use (for example within a boat rental company).

Figure 2.30   Refueling of the *Riviera 600* Frauscher boat is done in five minutes using a
standard 350 bar filler coupling plus simple exchange of an empty cartridge
for a full one.
(Reproduced from Frauscherboats.com, with permission.)

Figure 2.31   The Clean Power refueling station comprises an electricity power charger, hydrogen and payment units.
(Image courtesy of Fronius.)

For comparison, storing power in batteries over long periods of time
is linked to huge losses due to self-discharge (5–10% per month), while
the energy density is a fraction of that for hydrogen, which means that
storing energy in the summer in a battery of the same capacity would
mean that no energy was available in winter.

Figure 2.32 Made of passivated steel and 1 km in length, the first underground hydrogen pipeline in the world has been built in the Italian city of Arezzo and delivers pure H_2 at 3.5 bar to the fuel cells installed in four goldsmith companies.
(Reproduced from Ref. 54, with kind permission.)

Another example which shows that PV renewable hydrogen is far from being solely a research topic, is afforded by the world's first underground pipeline, which supplies H_2 to five customers in the Italian city of Arezzo, with a main channel of around 600 m (Figure 2.32).[54]

Solar hydrogen is generated by water electrolysis via PV modules over the roof of an off-grid building (*HydroLAb*, Figure 2.33), using four 5 kW hydrogen fuel cells and two 1 kW fuel cells to produce both electricity and heat for the whole building.[55]

Today, when solar panels generate more electricity than a home can use, the excess is simply fed back into the grid, essentially subtracting from the homeowner's utility bill. In an off-grid application, the excess is put into batteries, but fuel cells are more versatile and their price is declining rapidly.

Finally, an interesting approach to electrolytically generate hydrogen without the requirement for hydrogen storage has been introduced recently by Honda in the USA, with its Solar Hydrogen Station (Figure 2.34).[56] Users slowly refill their fuel cell vehicle overnight (an eight hour fill produces enough hydrogen, 0.5 kg, for a typical day's driving), by using

Figure 2.33   The HydroLAb is completely off-grid because PV solar panels provide electricity, and solar thermal vacuum tube panels provide heat for room heating and feed a solar cooling machine (5 kW, the smallest in Italy) in order to achieve air conditioning at zero emission in summer.
(Reproduced from Ref. 54, with kind permission.)

Figure 2.34   The Solar Hydrogen Station used to refuel Honda FCVs, installed in 2010 at the Los Angeles Center of Honda R&D Americas, employs the same 6.0 kW solar array thin film (CIGS) solar cells that power the system, with a compressor installed in 2001.
(Reproduced from Ref. 56, with kind permission.)

less expensive off-peak electrical power purchased by the grid at $0.05 kWh^{-1} or even less. During daytime peak power periods, the Solar Hydrogen Station exports high-valuepeak solar electricity to the grid, providing a cost benefit to the customer.

Designed as a single, integrated unit to fit in the user's garage, the new Honda hydrogen station reduces the size of the system, while the intuitive system layout enables the user easily to lift and remove the fuel hose, with no hose coiling when the hose is returned to the dispenser unit.

The previous solar hydrogen station system, studied by Honda in California since 2001, required both an electrolyzer and a separate compressor unit to create high pressure hydrogen. The compressor was the largest and most expensive component and reduced system efficiency. By creating a new high differential pressure electrolyzer, Honda engineers were able to eliminate the compressor entirely. This innovation also reduces the size of other key components to make the new station the world's most compact system, while improving system efficiency by more than 25% compared with the solar hydrogen station system it replaces.

One of the main messages of this book is that it has been the prolonged lack of hydrogen filling stations, both on large and small scales, that has caused the failure of hydrogen energy development in the last 20 years. Companies and local governments seem now to have understood this major gap and have started to deploy hydrogen refueling stations that use electrolyzes and renewable energy, namely PV or eolic electricity depending on the site location.

Figure 2.35, for example, shows the recent case of Hempstead, a town in New York state, where a 100 kW wind turbine provides the energy

Figure 2.35 The town of Hempstead, New York, located its Conservation and Waterways Department in Point Lookout; a 100 kW Northern Power 100 wind turbine was completed at Hempstead's hydrogen station. (Reproduced from greenfleetmagazine.com, with kind permission.)

necessary to create the hydrogen gas needed to power the town's fuel cell cars.[57]

Powered by winds off the Atlantic coast, the wind turbine is capable of generating up to 180 MWh of energy per year, providing an almost continuous source of low cost energy to split water. The resulting hydrogen fuel is dispensed from Long Island's *only* hydrogen fueling station, located adjacent to the turbine. The excess energy generated by the turbine will be fed into the town's grid, resulting in annual energy cost savings for local customers of the electric utility estimated at approximately $40 000. Overall, the wind turbine and hydrogen energy will save money, conserve natural resources, create jobs and provide smooth and pollution-free mobility to the town's citizens.

References

1. C. Russell, *Chem. World*, August 2003. www.rsc.org/chemistryworld/ Issues/2003/August/electrolysis.asp (last accessed on 04/01/2012).
2. C. R. Nave, Electrolysis of water, *HyperPhysics*, Georgia State University, Department of Physics and Astronomy, http:// hyperphysics.phy-astr.gsu.edu/hbase/thermo/electrol.html (last accessed on 04/01/2012).
3. F. T. Bacon, *Electrochim. Acta*, 1969, **14**, 569.
4. E. Guerrini and S. Trasatti, in *Catalysis for Sustainable Energy Production*, ed. C. Bianchini and P. Barbaro, Wiley-VCH, Weinheim, 2009, p. 235.
5. J. Divisek, J. Mergel and H. Schmitz, *Int. J. Hydrogen Ener.*, 1990, **15**, 105.
6. C. Neagu, H. Jansen, H. Gardeniers and M. Elwenspoek, *Mechatronics*, 2000, **10**, 571.
7. H. Wendt, in *Hydrogen as an Energy Carrier – technologies, systems, economy*, ed. C.-J. Winter and J. Nitsch, Springer, Berlin, 1988, p. 166.
8. E. Zoulias, E. Varkaraki, N. Lymberopoulos, C. N. Christodoulou and G. N. Karagiorgis, *A Review on Water Electrolysis*, http:// www.cres.gr/kape/publications/papers/dimosieyseis/ydrogen/A% 20REVIEW%20ON%20WATER%20ELECTROLYSIS.pdf (last accessed on 04/01/2012).
9. P. Heidebrecht and K. Sundmacher, in *Renewable Resources and Renewable Energy: A Global Challenge*, ed. P. Fornasiero and M. Graziani, CRC Press, Boca Raton, FL, 2nd edn, 2011.
10. I. C. Man, H.-Y. Su, F. Calle-Vallejo, H. A. Hansen, J. I. Martínez, N. G. Inoglu, J. Kitchin, T. F. Jaramillo, J. K. Nørskov and J. Rossmeisl, *ChemCatChem*, 2011, **3**, 1159.

11. C. Graves, S. D. Ebbesen, M. Mogensen and K. S. Lackner, *Renew. Sust. Energy Rev.*, 2011, **15**, 1.
12. H. Wendt and G. Kreisa, *Electrochemical Engineering*, Springer, Berlin, 1999.
13. R. Gammon, A. Roy, J. Barton and M. Little, *Hydrogen Renewables and Integration (HARI)*, case study report for the International Energy Agency, March 2006, http://ieahia.org/pdfs/ HARI.pdf (last accessed on 04/02/2012).
14. J. Ivy, *Summary of electrolytic hydrogen production: milestone completion report*, National Renewable Energy Laboratory, MP-560-35948. 28, 2004.
15. A. Konopka and D. Gregory, Hydrogen production by electrolysis: present and future, *10th Intersociety Energy Conversion Engineering Conference*, IEEE Cat. No. 75CHO 983-7 TAB, 1975.
16. Schatz Energy Research Center, Humboldt State University, *Development of a PEM Electrolyzer: Enabling Seasonal Storage of Renewable Energy*, California Energy Commission Energy Innovations Small Grant Program, 2005. www.energy.ca.gov/ 2005publications/CEC-500-2005-085/CEC-500-2005-085.PDF (last accessed on 04/01/2012).
17. P. Millet, N. Mbemba, S.A. Grigoriev, V.N. Fateev, A. Aukauloo and C. EtiÈvant, *Int. J. Hydrogen Energy*, 2011, **36**, 4134.
18. W. Hug, H. Bussmann and A. Brinner, *Int. J. Hydrogen Energy*, 1993, **18**, 973.
19. R. E. Clarke, S. Giddey, F. T. Ciacchi, S. P. S. Badwal, B. Paul and J. Andrews, *Int. J. Hydrogen Energy*, 2009, **34**, 2531.
20. A. Djafoura, M. Matouga, H. Bourasa, B. Bouchekimaa, M. S. Aidab and B. Azoui, *Int. J. Hydrogen Energy*, 2011, **36**, 4117.
21. B. Paul and J. Andrews, *Int. J. Hydrogen Energy*, 2008, **33**, 490.
22. R. García-Valverde, N. Espinosa and A. Urbina, *Int. J. Hydrogen Energy*, 2011, **36**, 10574.
23. G. E. Ahmad and E. T. El Shenawy, *Renewable Energy*, 2006, **31**, 1043.
24. A. Yilanci, I. Dincer and H. K. Ozturk, *Prog. Energy Combust. Sci.*, 2008, **35**, 231.
25. A. J. Bard and M. A. Fox, *Acc. Chem. Res.*, 1995, **28**, 141.
26. T. L. Gibson and N.A. Kelly, *Int. J. Hydrogen Energy*, 2010, **35**, 900.
27. N. Armaroli and V. Balzani, *Energy Environ. Sci.*, 2011, **4**, 3193.
28. M. Šúri, T. A. Huld, E. D. Dunlop and H. A. Ossenbrink, *Sol. Energy*, 2007, **81**, 1295.
29. E. Bilgen, *Sol. Energy*, 2004, **77**, 47.
30. http://data.worldbank.org/indicator/EP.PMP.SGAS.CD

31. US Department of Energy, *DOE Hydrogen Program: DOE H2A Analysis*, www.hydrogen.energy.gov/h2a_analysis.html (last accessed on 04/01/2012).

32. B. Kroposki, J. Levene,, K. Harrison, P. K. Sen and F. Novachek, *Electrolysis: Information and Opportunities for Electric Power Utilities*, National Renewable Energy Laboratory, Technical Report NREL/TP-581-40605, 2006.

33. C. Carpetis, *Int. J. Hydrogen Energy*, 1982, **7**, 287.

34. J. I. Levene, M. K. Mann, R.M. Margolis and A. Milbrandt, *Sol. Energy*, 2007, **81**, 773.

35. www.qsinano.com/apps_hgen.php (last accessed on 07/01/2012).

36. www.grid-shift.com (last accessed on 07/01/2012).

37. www.suncatalytix.com (last accessed on 07/01/2012).

38. J. J. Pijpers, M. T. Winkler, Y. Surendranath, T. Buonassisi and D. G. Nocera, *Proc Natl. Acad. Sci. USA*, 2011, **108**, 10056.

39. M. W. Kanan and D. G. Nocera, *Science*, 2008, **321**, 1072.

40. D. E. Hall, *Process of water electrolysis*, US Patent 4240887 (1979).

41. P. Nair, *Proc. Natl. Acad. Sci. USA*, 2012, **109**, 15.

42. ACTA's water electrolyser business 'ahead of expectations', *The Hydrogen Journal*, August 2 2011, www.thehydrogenjournal.com/displaynews.php?NewsID=676&PHPSESSID=k23b0o5an3sf5v8-fuvaj4kbiq3 (last accessed on 07/01/2012).

43. ITM Power, *HydroGen Results*, press release, 9 November 2011. www.itm-power.com/news/82/HydroGen + Results.html (last accessed on 07/01/2012).

44. www.rehydrogen.com (last accessed on 09/01/2012).

45. T. Graham Douglas and A. J. Cruden, D. Infield, A. Roy and P. Hall, *Int. J. Hydrogen Energy*, 2011, **36**, 7791.

46. www.theengineer.co.uk/sectors/energy-and-environment/news/british-firm-boosts-hydrogen-compression-and-storage/1011435.article#ixzz1l2hA7LFC (last accessed on 07/01/2012).

47. www.rabh2.co.uk (last accessed on 01/02/2012).

48. www.schatzlab.org/projects/real_world/schatz_solar.html (last accessed on 01/02/2012).

49. H. Steeb, W. Seeger and H. Aba Oud, *Int. J. Hydrogen Energy*, 1994, **19**, 683.

50. Renewables renewed, *SERC Energy News*, 2006, **1**, 4. www.schatzlab.org/docs/v1n4_dig_sm.pdf (last accessed on 01/02/2012).

51. *PHOTON International*, December 2011.

52. Hydrogenics successfully completes utility-scale grid stabilization trial with Ontario's independent electricity system operator, *Fuel-CellsWorks*, 16 June 2011. http://fuelcellsworks.com/news/2011/06/

16/hydrogenics-successfully-completes-utility-scale-grid-stabilization-trial-with-ontarios-independent-electricity-system-operator/ (last accessed on 02/02/2012).
53. www.zukunftsprojektwasserstoff.at/typo/fileadmin/user_upload/download/FAQ_fuel_cell.pdf (last accessed on 09/01/2012).
54. P. Fulini and E. Cecchini, Hydrogen project for Arezzo build up: an underground hydrogen pipeline testing fuel cells in industrial areas. Hydrolab – The link between hydrogen and renewable energies, *World Hydrogen Technologies Convention*, November 5, 2007, Montecatini Terme, Italy. http://old.caffescienza.it/download/Materiale/WHTC2007_Fulini-Cecchini.pdf (last accessed on 09/01/2012).
55. www.idrogenoarezzo.it (last accessed on 07/01/2012).
56. *Honda begins operation of new solar hydrogen*, press release, January 27, 2010. Stationhttp://world.honda.com/news/2010/c100127New-Solar-Hydrogen-Station/ (last accessed on 10/01/2012).
57. New York: new wind turbine powers hydrogen car fuel station, *FuelCellsWorks*, 25 January 2012. http://fuelcellsworks.com/news/2012/01/25/new-york-new-wind-turbine-powers-hydrogen-car-fuel-station/ (last accessed on 02/02/2012).

CHAPTER 3
Thermochemical Water Splitting

3.1 Concentrating Solar Power for Heavy Energy Demand

Solar energy technologies have the flexibility to address global power needs. Earth receives a vast amount of solar energy that is estimated to be approximately 120 000 TW (1 TW $= 10^{12}$ W), which vastly exceeds the current annual worldwide energy consumption rate of ~ 15 TW.[1] The latter figure includes all available forms of energy from electricity to gasoline combustion and is proportional with the population growth.

For example, energy consumption in 2010 increased by 5.6% compared to 2009.[2] Most of this power is currently produced by burning fossil fuels, namely coal, oil and natural gas (Table 3.1), which affords a clearly unsustainable situation given the limited reserves of these primary sources and the rapidly growing economies of large countries such as China, India and Brazil that undergo rapid industrialization. Moreover, the figures in Table 3.1 show a worrying growth in the consumption of coal, which is taking place not only in China and in India but also in the USA, countries where large reserves of coal exist and whose exploitation is increasingly pursued owing to the consistently high price of oil at $> \$90$ per barrel.

Therefore, when planning the transition towards a sustainable energy future we need a good understanding of the available, scalable and long-term solutions that can be applied globally.[1]

In other words, we must focus on solutions that meet not only our current energy demands but will have the potential to sufficiently cover

Solar Hydrogen: Fuel of the Future
Mario Pagliaro and Athanasios G. Konstandopoulos
© Mario Pagliaro and Athanasios G. Konstandopoulos 2012
Published by the Royal Society of Chemistry, www.rsc.org

Table 3.1 Fossil fuels and renewable sources: share of global energy consumption. (Adapted from Ref. 2, with kind permission.)

Oil's share of global energy consumption; rose by 3.1% over the year before	**34%**
Coal's share of global energy consumption; up by 7.6%, the highest since 1970	**29.6%**
Gas's share of global energy consumption; up by 7.4%, the biggest annual growth since 1984	**24%**
Share of renewables in global energy consumption	**1.8%**

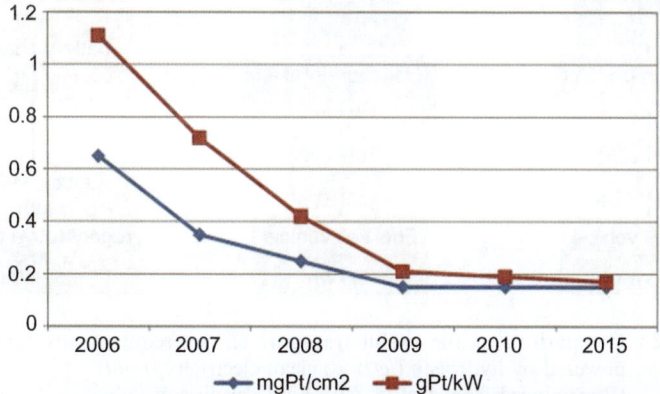

Figure 3.1 Trend of platinum load in PEM fuel cells, 2006–2015, according to the United States Department of Energy.
(Reproduced from Ref. 3, with kind permission.)

the world future demand. For example, current hydrogen fuel cells and lithium-based electric batteries for cars are *not* sustainable because the world reserves of both lithium (for batteries) and platinum[3] (for fuel cells) would be rapidly exhausted.

Assuming, with the US Department of Energy (DOE; Figure 3.1), that in 2015 the stack in fuel cells will use 0.2 g of platinum per kW, a fleet of 50 000 fuel cell vehicles (FCVs) with 80 kW stacks will demand 800 kg of platinum. Assuming that the current yearly output of General Motors, 2 980 000 cars, will be replaced with state-of-the-art FCVs, it would require 48 tonnes of platinum per year just for an automaker owning 4% of the world market in car production.

Given that the sun delivers 8000 times the present global power needs, it is rather safe to conclude that solar power is the *only* truly sustainable energy source. Sunlight of course is diluted: the yearly (average) solar power that reaches Earth's surface is about 170 W m^{-2}. Hence, when it comes to generating enough power to cover the escalating energy demands worldwide, we must *necessarily* focus on simple, low-tech

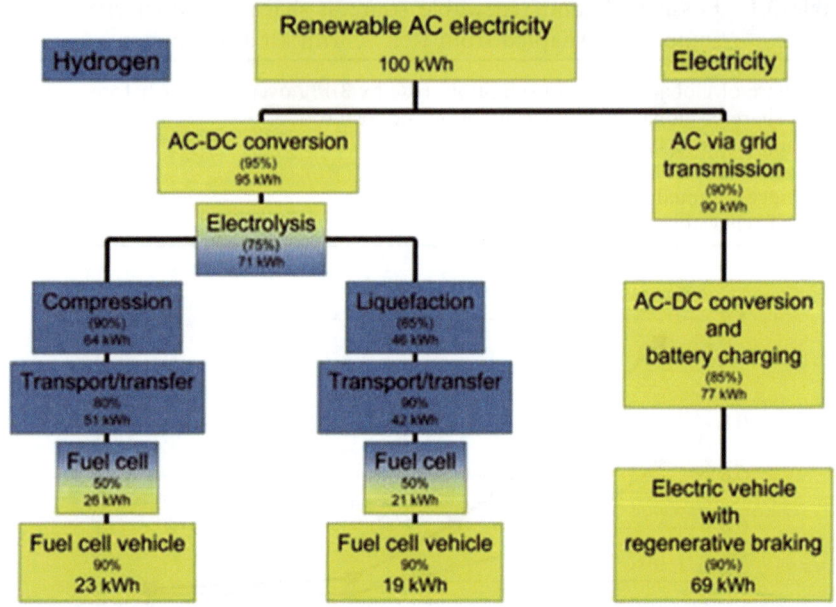

Figure 3.2 Comparison of the useful transport energy requirements for a vehicle
              powered by hydrogen (*left*) *vs.* clean electricity (*right*).
              (Reproduced from Ref. 5, with kind permission.)

solutions such as concentrating solar power (CSP) plants coupled to
energy storage in energy carriers, such as H_2, that can be made available
anywhere and at any time thanks to effective long-term storage of solar
energy.

Now, arguing that a hydrogen economy does *not* make sense, Bossel
showed in 2006 that if we assume standard production techniques for
hydrogen, the inefficiencies simply render the costs of storage and
transportation too high (Figure 3.2).[4] Emphasizing that hydrogen will
compete with its own source of energy, *i.e.* with electricity from the grid,
Bossel insists that we have to solve an *energy* problem, not an *energy
carrier* problem.[5] Low-tech CSP technology does indeed solve the
energy problem, while solar hydrogen addresses the inevitable demand
for effective storage of clean electricity. The CSP plants usually involve
solar thermal collection via parabolic mirrors, where focused sunlight
heats steam to about 600 °C to drive a turbine and generate electricity
(Figure 3.3).[6]

When we generate energy, the arrangement of matter becomes dis-
ordered, for example when steam is unavoidably heated to a high state
of disorder. Ordered structures such as nanostructures or crystalline
materials are not able to survive the unavoidable by-product of disorder

Figure 3.3   The PS-10 solar tower plant near Seville, Spain (courtesy of Abengoa
Solar). Solar energy is concentrated with heliostats to generate heat
for electricity generation. A similar concept can be applied to a plant for
solar hydrogen production.
(Reproduced from Wikipedia.org, with kind permission.)

when generating large quantities of energy. In other words, for heavy
energy demand a high-tech solution will never give both optimal reliability and efficiency.[1]

The CSP technology is both low-cost and low-tech and is capable of
producing large amounts of energy at a tiny fraction of the surface area
needed for photovoltaics (PV). Collection systems to concentrate solar
energy traditionally use parabolic reflectors (with trough, tower, dish
and, more recently, Fresnel's planar optical configurations) whose flux
concentration ratio C over a targeted area A at the focal plane, normalized with respect to the incident normal beam insolation I, is given
by Equation (3.1):[7]

$$C = \frac{Q_{solar}}{IA} \qquad (3.1)$$

Higher concentration ratios imply lower heat losses from smaller
areas and, consequently, higher attainable temperatures at the receiver.
When normalized to $I = 1000\,\mathrm{W\,m^{-2}}$ (the highest solar flux on a typical
sunny day), C is often expressed in units of "suns". The solar flux
concentration ratio typically obtained is at the level of 100, 1000, and
10 000 suns for trough, tower, and dish systems, respectively.

The first commercial CSP plants were erected in the 1980s in the
Californian Mojave Desert by the Israeli company Luz Industries
(Figure 3.4). These plants have a combined capacity of 354 MW and
today they generate enough electricity to meet the power needs of
approximately 500 000 people.[8]

Figure 3.4   Nine separate trough power plants, called Solar Energy Generating Systems (SEGS), were built in the 1980s in the Mojave Desert by the Israeli company Luz Industries. Synthetic oil captures this heat as the oil circulates through the pipe, reaching temperatures as high as 390 °C. (Reproduced from Wikipedia.org, with kind permission.)

Since then the technology has evolved considerably. For example, in modular CSP plants relatively small heliostats (movable mirrors) that use cheap linear Fresnel technology (costing between 50 and 60% of the costs of a parabolic collector per square meter)[9] track the sun and focus its energy onto tower-mounted receivers using non-toxic and readily available water as the unique thermal fluid (Figure 3.5).

The focused heat converts the fed water into superheated steam that drives a turbine generator to produce electricity. The steam passes through a steam condenser, reverts back to water through cooling, and the process repeats. Europe's first commercial solar concentrating power plant has operated smoothly since 2007, close to the Spanish city of Seville (see Figure 3.3). Overall, this CSP plant produces 11 MW of electricity, enough to power 6000 households. The plant currently operates with 624 heliostats, which concentrate the solar radiation on a thermal receiver located on a tower at a height of 115 m. The receiver converts the thermal energy into steam, which drives turbines that produce electricity.

Figure 3.5 Starting operation in October 2008, Ausra's Kimberlina 5 MW solar thermal power plant in Bakersfield was the first CSP plant to be built in the USA in nearly 20 years. Ausra was recently purchased by the French nuclear company Areva.
(Reproduced from Wikipedia.org, with kind permission.)

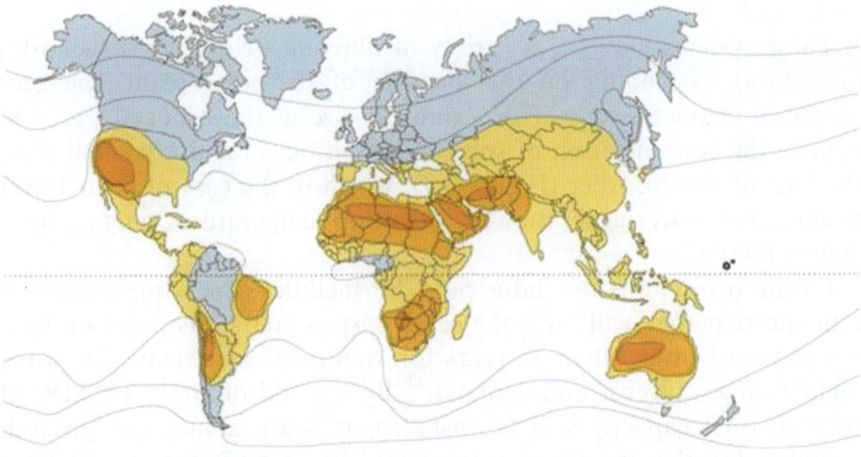

Figure 3.6 The Sun Belt is located roughly between the 40th parallels north and south, between southern Spain and South Africa, for example.
(Reproduced from www.solarmillennium.de, with kind permission.)

The best location for solar thermal power plants is the Earth's Sun Belt (Figure 3.6) because this is where the sun shines most frequently and where radiation is most intense. The CSP plants located within regions

Figure 3.7   Interconnected to the Californian grid, the 5 MW Sierra Sun Tower plant, built by eSolar in 2009, is the only commercial CSP tower facility in North America.
(Reproduced from esolar.com, with kind permission.)

of the Sun Belt have higher potential to store solar energy more efficiently, either as thermal power or by converting it into chemical fuels (solar fuels).[10]

These systems may use a variety of different field designs (heliostat allocation), depending on the location of the solar plant, the geographical characteristics of the land, the size of the heliostats, *etc.*, to control the concentration of solar radiation on a relatively small area, the face of the absorber (Figure 3.7), which in the case of solar tower facilities for power generation may develop temperatures ranging from 200 to 1000 °C.

Future projections for industrial CSP facilities generating electricity estimate that they will have the same cost as coal-, gas-, and oil-fired power plants in less than 15 years for "midload" electricity, *i.e.* in the middle ranges of cost and demand.[11] By the end of 2011, 850 MW of solar-thermal capacity will be installed in Spain alone, and ground-breaking for the construction of 2500 GW of CSP plants will have occurred in the USA. In total, project pipelines could represent 7000 MW of generating capacity worldwide. An overview of state-of-the art of technologies for solar thermal power production and fuel production has been published recently.[12]

Using some conservative assumptions and simple calculations, Abbott has shown that a total desert surface area of 500 km × 500 km can supply the whole world's energy needs.[1] If the world's power requirement is $P = 15\,\text{TW}$, then the solar farm footprint area is (Equation 3.2):

$$A = \frac{P}{I\eta_a\eta_g\eta_e\eta_l\eta_b} \tag{3.2}$$

Abbott assumed the employment of a solar Stirling dish of the type shown in Figure 3.8 but, of course, similar conclusions could be reached assuming the use of any other recent CSP technology.

The average insolation of a desert is conservatively set at $I = 300\,\text{W m}^{-2}$; $\eta_a = 0.54$ is the area fill factor of 10 m diameter dishes each occupying a plot of 12 m × 12 m to allow room for maintenance vehicles and cleaning equipment; $\eta_g = 0.3$ is the efficiency of the electricity production from a Stirling engine driven generator; $\eta_e = 0.5$ is the efficiency for electrolytically generating hydrogen; $\eta_l = 0.7$ is the efficiency to liquefy all the hydrogen; and $\eta_b = 0.5$ the efficiency of storage and transportation of hydrogen.

With all this taken into account, Equation (3.2) leads to an area of $1.76 \times 10^{12}\,\text{m}^2$, which is equivalent to a plot of size of 1330 km × 1330 km. In reality, with less pessimistic assumptions, the total area required equates to 500 km × 500 km only.[1]

As shown by Figure 3.6, many parts of the world's Sun Belt have hot desert regions ideal for hosting efficient solar CSP plants. Australia,

Figure 3.8   A Stirling solar dish, manufactured until 2011 by Stirling Energy Systems in Arizona.
(Reproduced from Wikipedia.org, with kind permission.)

China and the USA, for example, all have expansive stable dry deserts and could potentially supply power exceeding the whole world's energy needs. However, it will be far more economical in terms of energy distribution to have these solar farms widely distributed throughout the world. Solar dish farms around $4 \times 4 \, \text{km}^2$ in size are ideal for both economy of scale and wide distribution. This will also avoid the known geopolitical stresses caused by uneven distribution of oil in the world even if, given the higher sunlight requirements of CSP (compared with PV), the vast potential for energy generation by CSP will remain geographically unequally distributed relative to the main electricity consumers (located in Europe and in the USA).

3.2   Hydrosol: Thermochemical Water Splitting

Very high temperatures are required to dissociate water into hydrogen and oxygen. Given the thermodynamic restrictions, sufficient yields from the direct thermal splitting of water can only be achieved at temperatures above 2500 K. Temperatures this high impose extraordinary demands on materials and reactor design. Over the past 30 years numerous thermochemical cycles for hydrogen production through water splitting have been proposed and studied to a varying extent. Several cycles have been demonstrated at the laboratory scale, a couple have reached the pilot scale, but none has yet matured to production.

An interesting concept is that of oxide-based thermochemical cycles, during which a simple oxide (such as iron, zinc or cerium oxide) or a mixed oxide (such as a ferrite) cycles between a lower and a higher valence state, participating in an oxidation–reduction process that produces H_2 and O_2 in separate steps.[13] The concept has been proven experimentally for pairs of oxides of multivalent metals or metal–metal oxide systems (for example the ZnO–Zn system studied by Weimer, Steinfeld and co-workers, Figure 3.9).[14]

However, even though water splitting is taking place at temperatures lower than 700 °C, material regeneration (*i.e.* reduction) takes place at much higher temperatures (>1600 °C). In addition, despite basic research into active redox pairs, the solar reactors reported in the literature are based on particles fed into rotating cavity reactors, which are complicated and costly to operate. Hence, on the larger scale required to make solar-based H_2O splitting a practical technology in terms of quantity and cost using *only* the energy of the sun, an efficient and robust *redox material* is required to make the process operate at feasible temperatures.

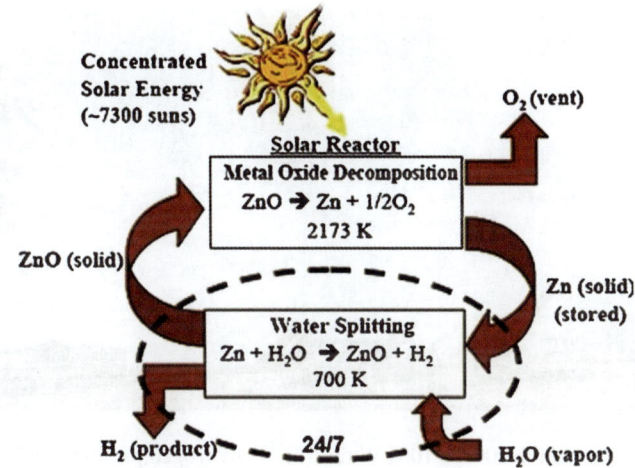

Figure 3.9 The ZnO–Zn thermochemical cycle. Primary issues revolve around both *i*) material development for operation day in and day out at 1800 °C in the presence of air and for rapid heating/cooling (*i.e.* thermal shock resistance) and *ii*) the development of methods to recover heat from the solar reactor while at the same time preventing recombination. (Reproduced from Wikipedia.org, with kind permission.)

Using a CSP plant it is easy to achieve temperatures in excess of 1200 °C through proper sizing and control of the heliostat field (a field of sun-tracking mirrors), hence a solar tower facility is the natural choice to host a solar thermochemical reactor.

Funded by the EU and coordinated by the Greek Aerosol and Particle Technology Laboratory of CPERI/CERTH (with partners the German Aerospace Center, DLR, the British company Johnson Matthey, the Spanish research center CIEMAT, and the Danish company Stobbe Tech), in 2008 the *Hydrosol-II* project established a 100 kW pilot plant at the Plataforma Solar de Almería in Spain (Figure 3.10).[15]

The thermochemical process – which involves an endothermic reaction that requires a significant energy input – employs a multichannel ceramic honeycomb reactor resembling that of the familiar catalytic converter of automobiles.

The reactor displayed in Figure 3.11 is coated with active mixed iron oxides (the redox material) with a high activity in the water splitting reaction. The reactor is thus heated by concentrated solar radiation using a set of heliostats that concentrate the solar energy into a certain area and lead to an increase in the reactor temperature (Figure 3.12).

Inaugurated on 31 March 2008, this solar reactor for the continuous production of solar hydrogen was the first ever closed solar-only,

Figure 3.10   In March 2008, a 100 kW reactor for hydrogen production through water splitting using solar energy was put into commission at the Plataforma Solar in Almería as part of the *Hydrosol* project. The reactor is located inside the tower on the right.
(Reproduced from hydrosol-project.org, with kind permission.)

thermochemical reactor for hydrogen production. Recently, the same research team has presented the results from the operation of the plant for 40 consecutive cycles of constant and continuous H_2 production in a two-day period.[16]

Significant concentrations of hydrogen were produced, with a conversion of steam of up to 30%. Operation has demonstrated that the combination of CSP facilities with high temperature processes will be a viable way to produce hydrogen at a reasonable cost without any greenhouse gas emissions, paving the way for a purely renewable *solar* hydrogen economy. Further scale-up of the technology and its effective coupling with CSP are in progress to demonstrate the large-scale feasibility of a solar hydrogen production plant.[17]

The design of this 100 kW pilot plant is based on a modular concept, and its scaling up to the megawatt range could follow both the traditional tower CSP approach (taller tower/larger heliostat field) or a parallel deployment of multiple units.

The scheme in Figure 3.12 shows that in the first step of water-splitting the activated redox reagent (usually the reduced state of a metal oxide) is oxidized by taking oxygen from water and producing hydrogen, according to the reaction in Equation (3.3).[18]

$$MO_{x-1} + H_2O(g) \rightarrow MO_{ox} + H_2$$ (3.3, exothermic)

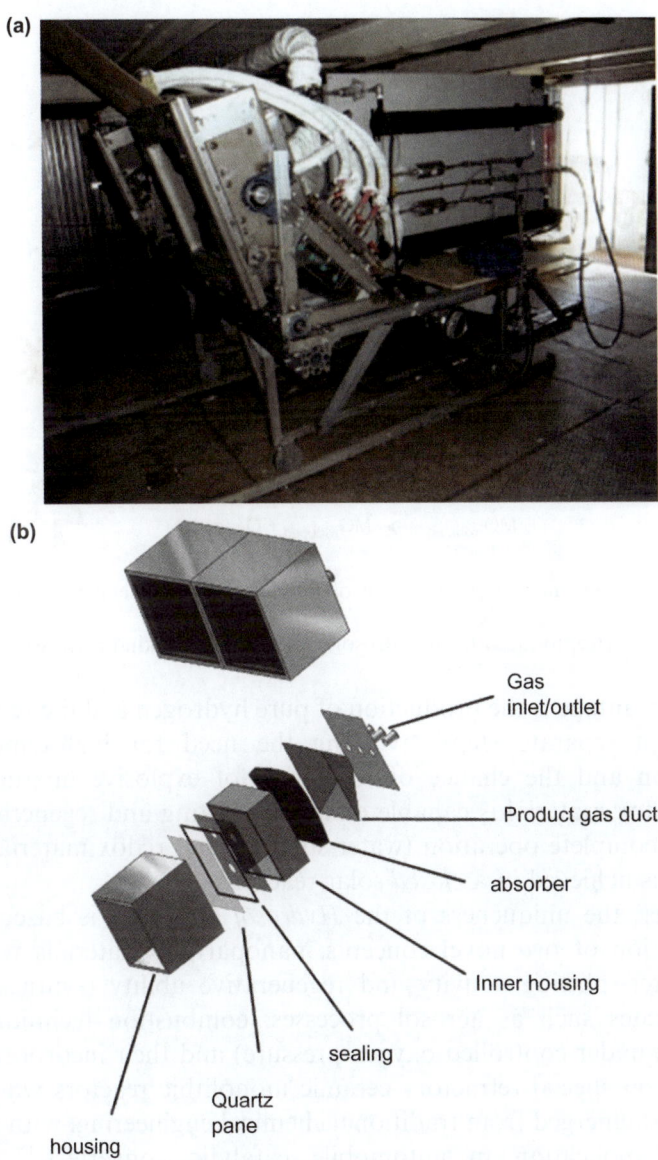

Figure 3.11 (a) The 100 kW *Hydrosol* pilot-scale reactor and (b) an exploded view of reactor design.
(Reproduced from Ref. 10, with kind permission.)

During the second step the oxidized state of the reagent is reduced, to be used again (regeneration), delivering some of the oxygen of its lattice according to Equation (3.4):

$$MO_{ox} \rightarrow MO_{x-1} + \tfrac{1}{2}O_2 \qquad\qquad (3.4, \text{endothermic})$$

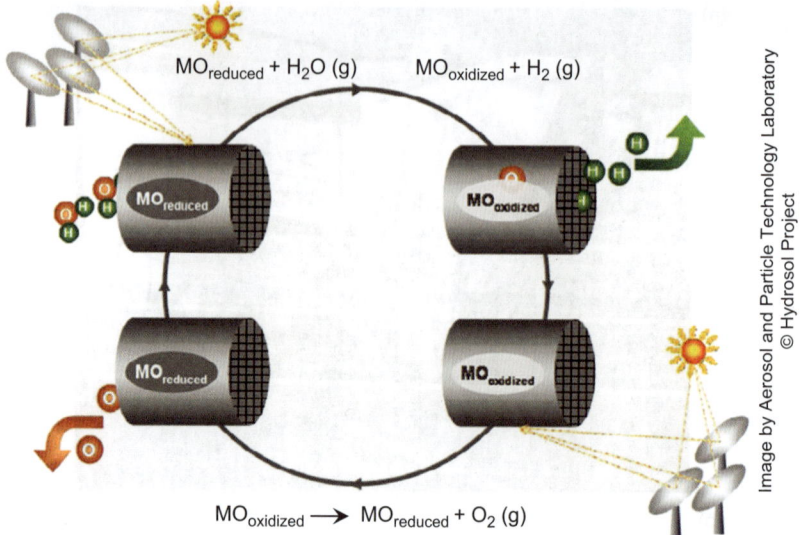

$MO_{reduced} + H_2O$ (g) $MO_{oxidized} + H_2$ (g)

$MO_{oxidized} \longrightarrow MO_{reduced} + O_2$ (g)

Image by Aerosol and Particle Technology Laboratory

© Hydrosol Project

Figure 3.12 Schematic representation of the solar thermochemical *Hydrosol* water splitting cycle.
(Reproduced from hydrosol-project.org, with kind permission.)

The advantage is the production of pure hydrogen and the removal of oxygen in separate steps, avoiding the need for high-temperature separation and the chance of formation of explosive mixtures. The active redox material is capable of water-splitting and regeneration, so that the complete operation (water-splitting and redox material regeneration) is achieved in a *closed* solar reactor.

In brief, the uniqueness of the *Hydrosol* approach is based on the combination of *two* novel concepts: nanoparticle materials with very high water-splitting activity and regenerative ability (synthesized by novel routes such as aerosol processes, combustion techniques and reactions under controlled oxygen pressure) and their incorporation as coatings on special refractory ceramic monolithic reactors whose geometry first emerged from traditional chemical engineering, with its most familiar application in automobile catalytic converters.[19] Coated monolithic reactors are therefore one of the two *enabling* technologies for renewable solar hydrogen production.

The solar thermochemical reactor for the production of hydrogen from water-splitting is constructed from special refractory ceramic thin-walled, multi-channeled (honeycomb) monoliths (Figure 3.13) that absorb solar radiation.

The reactor contains no moving parts, and converts the solar radiation into hydrogen very efficiently.[20] When steam passes through the

Figure 3.13 The monolith channels are coated with active water-splitting materials capable of splitting the steam passing through the reactor by 'trapping' its oxygen and leaving, in the effluent gas stream, pure hydrogen as the product.
(Reproduced from hydrosol-project.org, with kind permission.)

reactor, the coating material splits the water molecules by adsorbing and incorporating oxygen to form a higher valence state oxide. The effluent gas stream then consists of pure H_2.

The temperature in the reactor is then increased, for example by focusing more mirrors onto the aperture of this reactor. The feed gas stream is cut off, the trapped oxygen is released and the active coating is regenerated. Two reaction chambers (designated as Eastern and Western modules) are operating in parallel, one for water splitting and one for regeneration.

Accurate temperature control is necessary in particular for the high temperature reaction, the regeneration, on the one hand to avoid overheating and on the other hand to ensure sufficient reaction rates.

Figure 3.14 demonstrates the effect of varying the number of different heliostats focused on the two apertures and the feasibility of the described control concepts when using only the number of heliostats for temperature control. For both temperatures, 800 °C and 1200 °C, the sufficient control can be applied using the heliostats to ensure steady states.

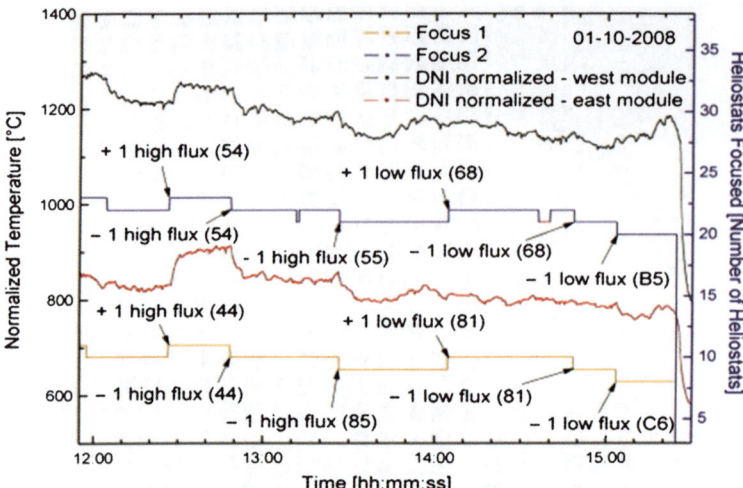

Figure 3.14 Effect of a varying number of heliostats on temperature in the *Hydrosol* reactor.
(Reproduced from Ref. 16, with kind permission.)

One of the prototype redox materials for this kind of reaction is Fe_3O_4. The redox pair in this case is FeO–Fe_3O_4. In practice, the pure oxide cannot be cycled because the temperature needed to reduce Fe_3O_4 thermally is higher than its melting point. However, replacing some of the iron in Fe_3O_4 with other metals, such as zinc, manganese, nickel or cobalt, can lower the reduction temperature while maintaining the spinel structure of these ferrite materials. Integrating the ferrites into a stabilizing matrix, such as yttrium-stabilized zirconia or cerium oxide (ceria), finally, slows down sintering and deactivation of the metal oxide.[21]

In the *Hydrosol 2* pilot plant, the composition of the off-gas stream was detected by a gas chromatograph (GC). A plot of hydrogen concentrations in the product stream of the plant is displayed in Figure 3.15. The first broad peak at $t = 5000$ s is attributed to the splitting of residual water in the apparatus.

Apparently the highest output of hydrogen was produced during the first cycle at about $t = 7000$ s. The measured concentration corresponds to a conversion of 30% of the steam fed in. After that, a reduction of hydrogen concentration, and therefore of the yield, by a factor of about two was observed.

This effect is similar to what has been observed earlier in smaller reactors in the laboratory and in the solar furnace, and is attributed mainly to deactivation of the particular redox system (and to a minor extent to inhomogeneous temperature distribution of the absorber).

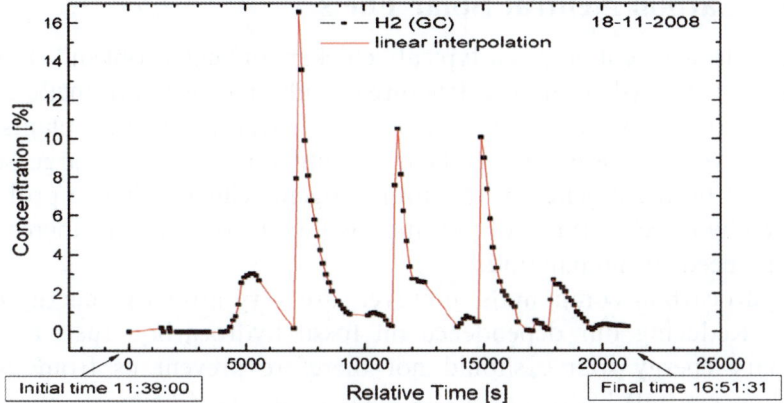

Figure 3.15 Concentration of hydrogen in product stream in the *Hydrosol* 2 pilot plant reactor.
(Reproduced from Ref. 16, with kind permission.)

There is evidence that some of the zinc in the particular ferrite formulation volatilizes during cyclic operation, resulting in a reduction of the activity of the redox material from its initial value. The strongly diminished hydrogen production indicated by the last peak was caused by the occurrence of a leakage and therefore by air infiltration into the reactor. Experiments with new, more robust ferrite compositions are expected to start in the spring of 2012.

Konstandopoulos and colleagues are now working to scale up their technology and build a 1 MW hydrogen-producing plant, in a project known as *Hydrosol 3D*, which involves the pre-design and design of the whole plant, including the solar hydrogen reactor and all necessary upstream and downstream units needed to feed in the reactants and separate and bottle the products.

The Hydrosol-3D consortium consists of the Aerosol & Particle Technology Laboratory of CPERI/CERTH, Germany's DLR, Spain's CIEMAT, and finally the French company Total and the Belgian company Hygear.[22] Two alternative options are currently being analyzed: adapting the hydrogen production plant to an already available solar facility or developing a new, completely optimized hydrogen production/solar plant.

These and related developments and large scale demonstration are now urgently needed because, as stated by Steinfeld,[7] the weaknesses in the economic evaluations of thermochemical solar hydrogen production are related primarily to the *uncertainties* in the viable *efficiencies* and *investment costs* of the various components due to their early stage of development and their economy of scale.

3.3   Carbon Neutral Solar Fuels

The hydrogen economy is a typical "chicken and egg problem" (Figure 3.16). Until a hydrogen infrastructure is built, hydrogen production will not reward investment. Hence, (solar) hydrogen production plants, as well as vehicles, are not yet being manufactured on the large scale required by the urgency of the anthropogenic climate change problem caused by three centuries of burning fossil fuels to power the increasing energy needs of humankind.

Hydrocarbon compounds, however, are very attractive energy carriers. Reducing our dependence on fossil hydrocarbon fuels as our primary energy source should not therefore prevent us from using (carbon neutral) hydrocarbons as energy carriers. Chemicals offer the advantages of being transportable as well as being able to be stored for extended periods of time. This point is important because energy demand is rarely synchronous with or geographically matched to incident solar radiation.

In general, solar thermochemical processes for the production of synthetic fuels using concentrated solar radiation are thermodynamically favored because they inherently operate at high temperatures and utilize the entire solar spectrum. Recently, therefore, Konstandopoulos and his colleagues have modified the Hydrosol technology successfully and have managed, with a similar reactor technology, to produce carbon monoxide by splitting of carbon dioxide (CO_2).[23]

Indeed, when CO_2 is passed through the Hydrosol reactor, the coating material splits the molecules by adsorbing and incorporating oxygen to form a higher oxide. The effluent gas stream then consists of pure CO. The temperature in the reactor is increased subsequently by focusing more

Figure 3.16   Which will come first: Hydrogen infrastructure or hydrogen production?

mirrors onto the aperture of the reactor and the feed gas stream is cut off, which releases the trapped oxygen and regenerates the active coating.

If the operation of the CO_2 splitting reactor is "combined" with the operation of the Hydrosol reactor, the carbon monoxide and hydrogen produced simultaneously will react to give synthetic fuel, produced either by the well-known Sabatier or by the Fischer–Tropsch process to convert H_2 and CO into liquid hydrocarbons for the transportation sector, or into polymers.

In the Sabatier process the two gases (CO and H_2, known as "synthesis gas") are heated at high pressure in the presence of a nickel catalyst to produce methane or methanol; in the Fischer–Tropsch process an iron-based catalyst is used to generate liquid hydrocarbon fuels. In addition to solar fuels, the solar "synthesis gas" can be effectively employed to synthesize a wide variety of hydrocarbon polymers ("solar plastics"), contributing further to a sustainable future (Figure 3.17).

These processes offer a very good alternative for dealing with the problem of carbon storage. The CO_2 captured from power plants could constitute an ideal raw material for the production of synthetic fuels, rather than being buried in underground storage sinks. Such a development permits the continued use of the existing hydrocarbon fuel infrastructure to distribute carbon neutral solar fuels (hydrogen and hydrocarbons). In this way a viable and sustainable solution to problems of both hydrogen storage and carbon storage can be provided.

A vision of a biomimetic carbon neutral fuel grid for Europe is given in Figure 3.18, where CO_2 from the north is conveyed by pipelines

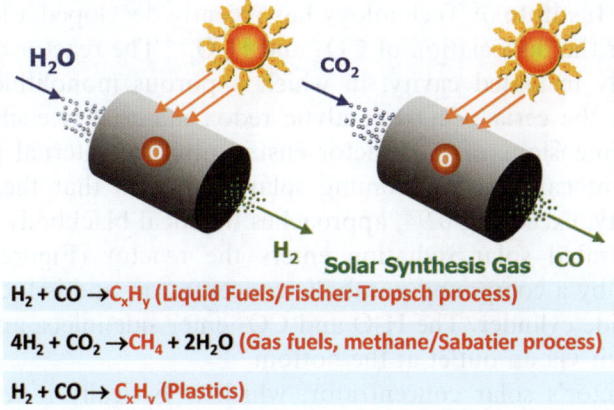

Figure 3.17 Coupling solar hydrogen generation with CO_2 splitting could produce green energy and polymers from sunlight, H_2O and CO_2, similar to what Nature does to synthesize organic matter.

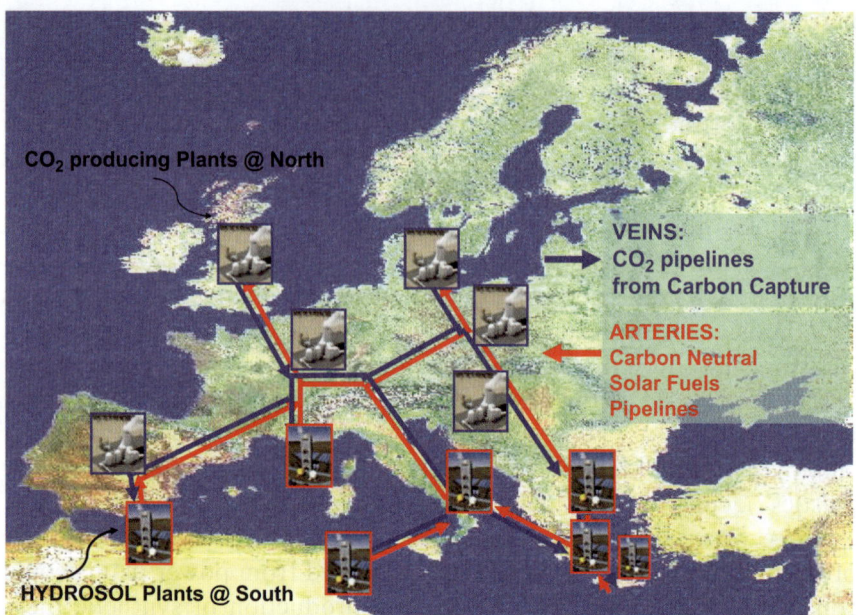

Figure 3.18   A biomimetic system for sustainable energy as conceived by Hydrosol project researchers.

("veins") to the sunny south, where, through Hydrosol plants, it is converted into solar hydrocarbons and these are distributed to the areas of demand by solar fuel pipelines ("arteries").[24]

A similar approach is being pursued by Aldo Steinfeld and his team at the Swiss Federal Institute of Technology, ETH Zurich. A research team consisting of ETH, the Paul Scherrer Institute (PSI), and the California Institute of Technology has recently developed a laboratory reactor for the dissociation of CO_2 and H_2O.[25] The reactor consists of a thermally insulated cavity, in which a porous monolithic cylinder containing the ceria (CeO_2) catalytic redox material is enclosed. The selected dimensions of the reactor ensure multiple internal reflections and efficient capture of incoming solar energy so that the apparent absorptivity, exceeding 0.94, approaches the ideal blackbody limit.

Concentrated solar radiation enters the reactor (Figure 3.19), is intensified by a compound parabolic concentrator, and is focused on a cerium oxide cylinder. The H_2O and CO_2 enter side inlets, and O_2, H_2, and CO exit via an outlet at the bottom.

The reactor's solar concentrator, which is basically a set of giant curved mirrors that gather sunlight from a wide area, is the most difficult part to build. The redox pair CeO_2–Ce_2O_3 is the other main redox system for thermochemical water splitting. Ceria has the advantage that

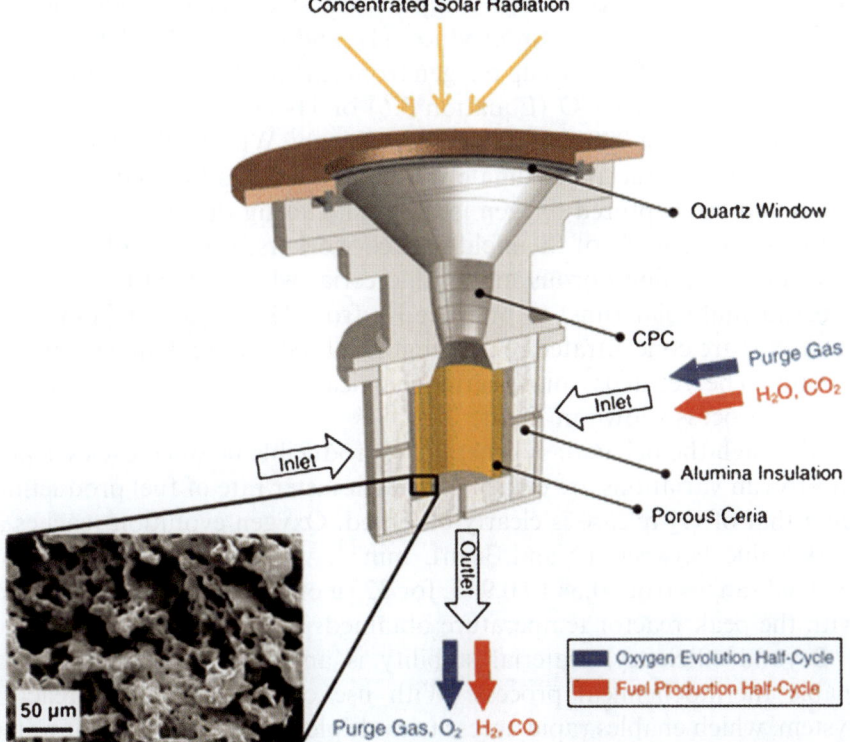

Figure 3.19 The solar reactor for the two-step, solar-driven thermochemical production of fuels developed by Steinfeld and co-workers consists of a thermally insulated cavity receiver containing a porous monolithic ceria cylinder.
(Reproduced from Ref. 25, with kind permission.)

the melting point is higher than the temperature required for the thermal reduction step. Moreover, ceria is a highly attractive choice for two-step thermochemical cycling because it displays rapid fuel production kinetics and high selectivity, owing to the absence of distinct oxidized and reduced phases.

The two-step H_2O/CO_2 splitting solar thermochemical cycle based on oxygen-deficient ceria is represented by Equations (3.5 to 3.7):

$$\text{High-T reduction: } CeO_2 \xrightarrow{\Delta H} CeO_{2-\delta} + \delta/2 O_2 \qquad (3.5)$$

$$\text{Low-T oxidation with } H_2O: CeO_{2-\delta} + \delta H_2O \rightarrow CeO_2 + \delta H_2 \qquad (3.6)$$

$$\text{High-T oxidation with } CO_2: CeO_{2-\delta} + \delta CO_2 \rightarrow CeO_2 + \delta CO \qquad (3.7)$$

In the first high-temperature step, the ceria is thermally reduced to a non-stoichiometric state (T > 1673 K) and oxygen is released

(Equation 3.5). In the lower temperature steps, ceria is re-oxidized with H_2O and/or CO_2 to produce H_2 and/or CO. In detail, non-stoichiometric CeO_2 takes up oxygen from carbon dioxide or from water and produces either CO (Equation 3.6) or H_2 (Equation 3.7), at temperature ranges of 700 and 500 °C, respectively. With further increase of the temperature (at approximately 1500 °C), ceria is thermally reduced again and the captured oxygen is released, closing the cycle.

In the experiment of Steinfeld and co-workers, using a solar cavity-receiver containing porous monolithic ceria, which aimed (in separate experimental solar runs) to produce H_2 from H_2O and CO from CO_2, solar rays are concentrated to a strength of 1500 suns and directed into a reactor. The result is solar-thermochemical fuel production from the cycling process shown in Figure 3.20.

Although the behavior is generally reproducible between cycles, some run-to-run variations are evident. A much faster rate of fuel production than that of O_2 release is clearly observed. Oxygen evolution reaches a peak value between 17 and 34 mL min^{-1}, whereas the total amount evolved ranges from 0.54 to 0.94 L for 325 g of ceria, which is correlated with the peak reactor temperature obtained.

Beyond efficiency, material stability is an essential criterion for a viable thermochemical process. With use of the differential reactor system, which enables rapid access to multiple cycles, 500 cycles of water dissociation were performed without interruption.

The results indicate that, after an initial stabilization period of about 100 cycles, both the oxygen and hydrogen evolution rates remained essentially constant for a subsequent 400 cycles (Figure 3.21).

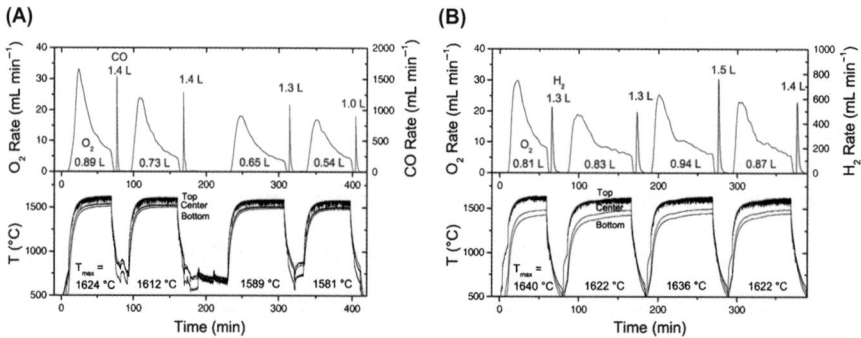

Figure 3.20   Thermochemical cycling of ceria (325 g) using the solar reactor with (A) CO_2 and (B) H_2O as oxidant, showing the oxygen and fuel evolution rates as well as the total volume of gas evolved.
(Reproduced from Ref. 25, with kind permission.)

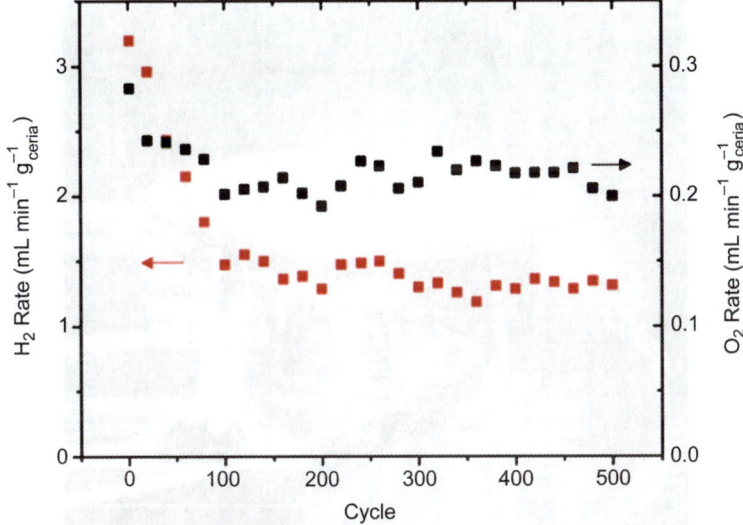

Figure 3.21 The O_2 (black) and H_2 (red) evolution rates for 500 water-splitting cycles. CeO_2 was cycled between 1500 °C and 800 °C. (Reproduced from Ref. 25, with kind permission.)

Scanning electron microscopy examination of samples of porous ceria that underwent heat treatment under similar conditions revealed that the decrease in reaction rate was accompanied by an increase in particle size. The morphology stabilized after 24 hours of heat treatment at 1500 °C, much as the fuel production rate stabilized after an initial period.

The originally reported solar-to-fuel efficiencies of 0.7 to 0.8% are largely limited by the system scale and design (Figure 3.22) rather than by chemistry. For comparison only, the solar-to-fuel energy conversion efficiency obtained in this work for CO_2 dissociation is about two orders of magnitude greater than that observed with state-of-the-art (in 2010) photocatalytic approaches.[27]

A thermodynamic analysis[28] indicates that efficiencies of 16% or more are achievable with the new reactor. Hence, the team optimized the solar reactor prototypes (at the 10 kW power level) for maximum solar-to-fuel energy conversion efficiency, and is currently scaling-up the system for industrial applications (at the MW power level) using concentrating solar tower technology.

Most recently, Steinfeld has demonstrated that the same geometrical cavity-type configuration reactor, this time packed with porous ceria felt, can co-produce H_2 and CO (syngas) by *simultaneously* splitting a

Figure 3.22   The reaction chamber in which sunlight becomes chemical energy. This picture shows the reaction chamber of the new solar collector illuminated by light coming from a solar simulator. A quartz window at the top allows both infrared and ultraviolet radiation to enter the chamber in which the cerium oxide is deposited.
(Reproduced from Ref. 26, with kind permission.)

mixture of H_2O and CO_2.[29] In detail, ten consecutive H_2O/CO_2 gas splitting cycles have been performed over eight hours with a 3 kW solar cavity receiver-reactor containing porous ceria felt exposed directly to high-flux (>2800 suns) thermal radiation.

A constant and stable syngas composition (Figure 3.23), showing stable fuel production, was observed, which demonstrates the feasibility of using ceria-based redox cycles to produce repetitive and controlled amounts of syngas in a solar reactor that closely replicates the conditions expected in practical solar fuel applications.

Indeed, the solar reactor design is simple and robust, affording clear benefits – simplicity, robustness, stability, and use of earth-abundant elements – that render this thermochemical approach feasible for large-scale implementation. The material stability over 500 thermochemical cycles observed with monolithic ceria in the separate generation of hydrogen and carbon monoxide is already suitable for realistic applications. Furthermore, the abundance of cerium, which is comparable to

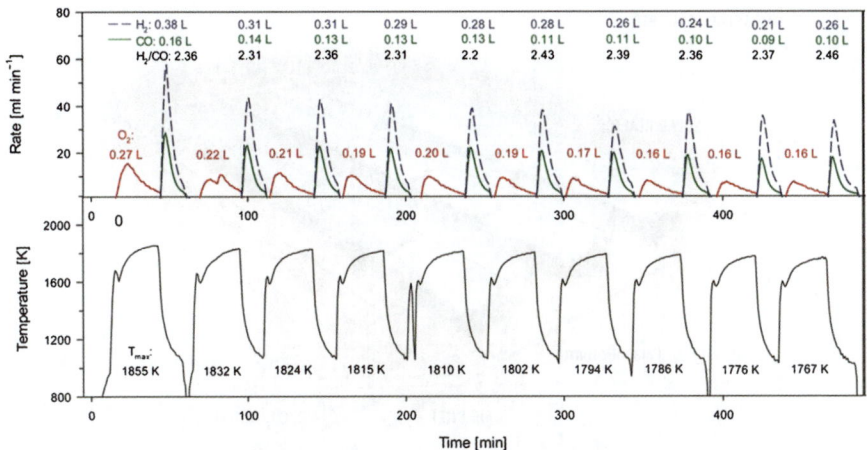

Figure 3.23 Temperature of the ceria felt, gas production rates, total amount of evolved gases, and $H_2 : CO$ molar ratios during ten consecutive splitting cycles.
(Reproduced from Ref. 29, with kind permission.)

that of copper,[30] is such that the approach is applicable at scales relevant to global energy consumption.[31]

The decision was taken by ETH not to patent the new discovery to enable further research.[32] Research is now focused on finding proper dopants for ceria-based materials. The already investigated dopants (*e.g.* Gd and Sm) affect the thermodynamics of the reduction of ceria and as a result are expected to increase its reduction degree at lower temperatures. In that way the overall lifetime of the materials and the reactor will be extended.

Carbon dioxide can of course be accumulated from the atmosphere. Steinfeld and co-workers have developed another reactor made of a transparent tube filled with pellets of CaO (calcium oxide). As the light heats the tube and brings its contents to 400 °C, air mixed with a small amount of steam is pumped in at the bottom and up through the pellets. At this temperature, CaO reacts with CO_2 to form calcium carbonate ($CaCO_3$) and in less than 15 minutes removes all carbon dioxide in the air, decreasing it from 385 parts per million to practically zero.[33] Subsequently, the intake valve is closed and the temperature in the reactor is raised to 800 °C by intensifying the light, which causes the $CaCO_3$ to release the CO_2 as a stream of pure gas and converts the calcium carbonate back into calcium oxide. The reactor was taken through five cycles of absorption and release with no decline in performance.

Steinfeld and his team have thus developed a system that uses atmospheric CO_2 to feed the solar fuel process (Figure 3.24). A parabolic

THE SOLAR SCRUBBER

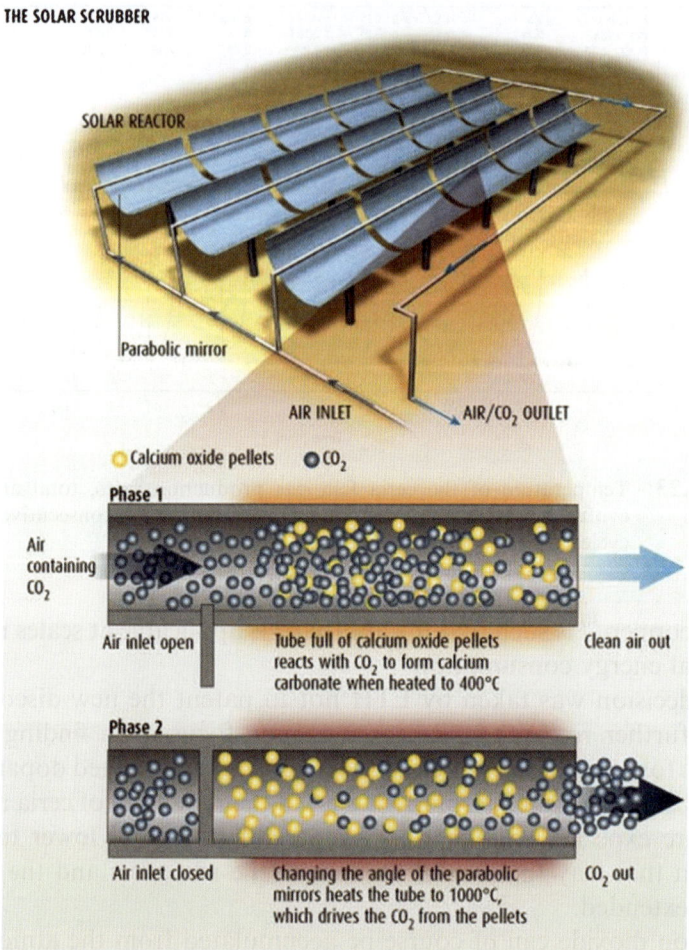

Figure 3.24   The O_2 (black) and H_2 (red) evolution rates for 500 water-splitting cycles. CeO_2 was cycled between 1500 °C ($pO_2 = 10-5$ atm, flow rate = 3.2 L min^{-1} g^{-1} of ceria, 10 min, ramp rate = 100 °C min^{-1}) and 800 °C ($pH_2O = 0.13$–0.15 atm, flow rate = 0.75–0.76 L min^{-1} g^{-1} of ceria, 10 min).
(Reproduced from http://nextbigfuture.com/2009/01/co2-capture-from-air-for-fuel-or.html, with kind permission.)

mirror concentrates solar light onto a chamber containing calcium oxide (CaO). When calcium carbonate is heated to 800 °C it releases a pure stream of CO_2 that is fed into a second reactor, in which a solar concentrator heats zinc oxide to 1700 °C, causing it to release oxygen molecules, leaving metallic zinc. The temperature is then lowered and CO_2 and steam are pumped in, which react with the pure Zn to form syngas.

Finally, a team led by James Miller at the US Sandia National Laboratories have also built a solar reactor based on a counter-rotating-ring receiver/reactor/recuperator concept (termed the Counter Rotating Ring Receiver Reactor Recuperator, *CR5*), for the production of solar fuels from hydrogen (obtained through water splitting) and carbon monoxide (obtained by CO_2 splitting), by making use of the two-step ferrite or ceria cycles.[34] The most recent *CR5* consists of two chambers separated by rotating rings of cerium oxide (Figure 3.25). As the rings spin, a large parabolic mirror concentrates solar energy onto one side, heating it to 1500 °C and causing the ceria there to release oxygen gas into one of the chambers, from which it is pumped away.

As the ring rotates further it cools before it swings round to the other chamber where CO_2 is pumped, causing the cooled, non-stoichiometric ceria to split carbon dioxide and produce carbon monoxide. The process also works with water instead of CO_2, with the reaction this time producing hydrogen. Initial test results for the *CR5* prototype, in the 16 kW National Solar Thermal Test Facility (NSTTF) solar furnace in Albuquerque, demonstrated that the process can produce carbon monoxide, although the failure of certain parts meant that the device did not operate continuously for more than a few seconds at a time.[36]

The team is now working to improve reliability while building a bigger reactor with 28 rotating rings to process more CO_2 and water. The short-term goal for the *CR5* prototype is to demonstrate a solar to chemical conversion efficiency of at least 2%. To achieve the overall

Figure 3.25 The Sandia Counter Rotating Ring Receiver Reactor Recuperator, CR5. (Reproduced from Ref. 35, with kind permission.)

long-term goal of 10% efficient conversion of sunlight to petroleum, the thermochemical solar conversion of sunlight to CO needs to be at least 20% efficient.

In conclusion, the results obtained by these teams in the last decade provide evidence for the viability of CSP-based thermochemical approaches to solar hydrogen and solar fuel generation. We believe that this technology will soon evolve into a central energy technology in our common, sustainable future. Eventually, solar H_2 is considered to be a feasible (and ultimate) emissions-free solution. However, carbon neutral solar fuels, obtained from the solar conversion of water and recycled CO_2, could cover the intermediate period during the passage from conventional to renewable hydrogen energy. This will give the time required to solve the issues related to hydrogen handling, storage, transport and distribution infrastructure that are challenging for its implementation in the near future.[37]

3.4   Solar Hydrogen and the Electron Economy

According to the US National Academy of Engineering, the electric grid was the most significant engineering achievement of the 20th century.[38] The interconnected series of transmission wires, metal towers, voltage converting substations and their associated control structure that make up the electric grid is of course existing infrastructure of immense value (Figure 3.26). It will be used increasingly in the solar hydrogen economy, in which a continuous source of power – solar hydrogen accumulated during the day – will be burned (or oxidized in fuel cells) to meet customers' demands and provide power at night and on cloudy days, as well as when demand peaks.

In other words, once generated in CSP hydrogen plants, electricity and *not* compressed hydrogen will be sent directly to the appliances using the grid (an existing infrastructure), exploiting the intrinsically higher efficiency of what Bossel has called the "electron economy". In Bossel's words:[5]

An electron economy can offer the shortest, most efficient and most economical way of transporting the sustainable 'green' energy to the consumer. Electricity could provide power for cars, comfortable temperature in buildings, heat, light, communication, etc.

In a sustainable energy future, electricity will become the prime energy carrier. We now have to focus our research on electricity storage, electric cars and the modernization of the existing electricity infrastructure.

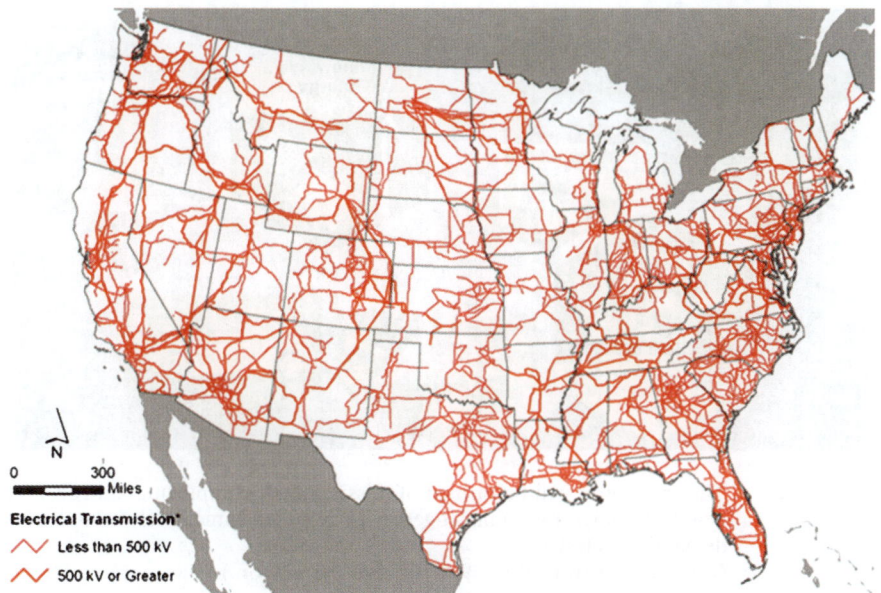

Figure 3.26 A scheme of the US grid in 2009.
(Reproduced from Ref. 38, with kind permission.)

Solar hydrogen is, indeed, an excellent way to store electricity. Even though the efficiency of the solar hydrogen economy will be affected by wastage in the two conversions – from physical to chemical and from chemical to physical energy – it will be sufficient to deploy more solar collectors and low cost electrolyzers in order to obtain the required amounts of hydrogen to power the world's current (15 TW) and future power needs.

In other words, a *solar* hydrogen economy is based on two such conversions (electrolysis and fuel cells or hydrogen engines) along with electricity generation from a free fuel (sunlight), whose immense abundance will allow us simply to overcome the efficiency limitations of this double conversion of energy (physical → chemical → physical).

Moreover, in the "distributed generation" scheme in which ever more grid homes and industrial customers are contributing, by producing some or all of their own power, solar hydrogen will help us to face the changes required to adapt the grid to allow more flexible options for future electricity production, transmission and distribution.[39]

Increased use of solar energy generation means more variable power supplies that have to be backed up by storage or by other power generation systems. Demand response for both large and small consumers is

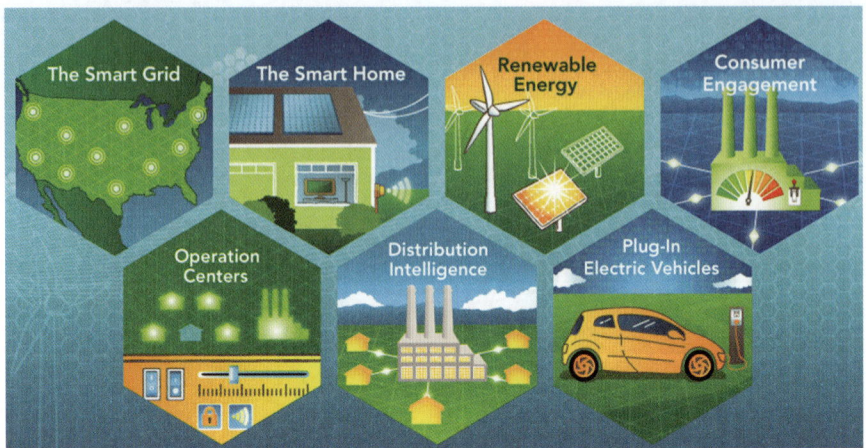

Figure 3.27 The smart grid will be formed of mission lines, equipment, controls and new technologies working together to respond immediately to our new demand for electricity.
(Reproduced from Ref. 39, with kind permission.)

becoming an increasingly viable way to deal with peak energy use. Advances in distributed generation, such as home solar panels, both assist and are assisted by the creation of the smart grid (Figure 3.27). The load production is closer to where it is needed and can help to balance peak load needs.

For example, in Canada, BC Hydro converts surplus hydroelectric power generated at its Clayton Falls hydroelectric plant in Bella Coola, British Columbia, into hydrogen, stores it and then uses it in a 100 kW fuel cell to provide power as needed.

Hydro power is already a good storage system for electrical energy, because utilities can postpone energy production by pumping the water backwards in the dam, with good efficiency.[40] In the case shown below the hydro power is actually micro-hydro power, with no storage for water, namely it is a "run of river" system (Figure 3.28) which does not provide the typical output control a traditional dam provides.

In the Clayton Falls Hydrogen Assisted Renewable Power System (HARP) system, two methods are used to store the electricity. In the first method hydrogen is produced through electrolysis, and is then stored as a gas in high pressure tanks. The second method uses an electrochemical regenerative fuel cell, known as the flow battery, to store the energy. A microgrid controller manages the power system (Figure 3.29) by monitoring supply and demand and determining when

Figure 3.28   Clayton Falls Hydroelectric Generating Station was built in 1962 to reduce demand for diesel fuel. This is a run-of-the-river facility that requires no storage of water in the Clayton Falls headpond. (Reproduced from Ref. 41, with kind permission.)

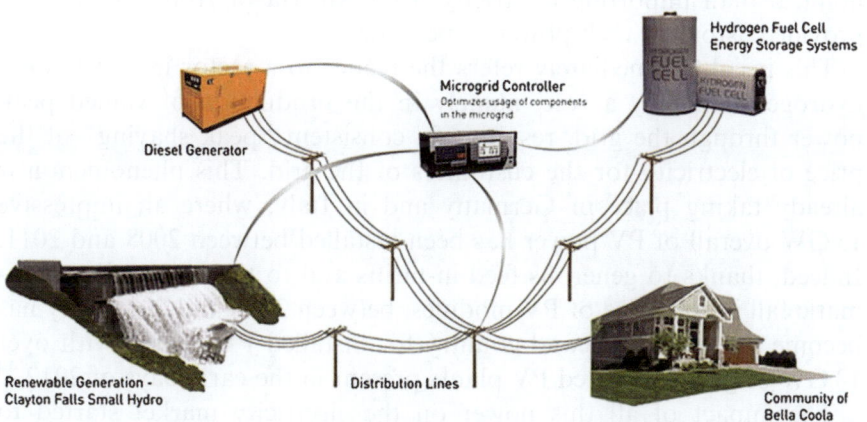

Figure 3.29   The Hydrogen Assisted Renewable Power (HARP) system. A schematic of microgrid system from Bella Coola, British Columbia. (Reproduced from Ref. 42, with kind permission.)

to convert power into hydrogen and when to generate power from hydrogen to meet spikes in demand.

During peak periods the stored hydrogen (Figure 3.30) is fed into a 100 kW fuel cell to generate electricity. At the same time, the flow battery produces 100 kW of electricity directly to the community. Together these two generators in the HARP system reduce annual diesel consumption by 200 000 L.[42]

Figure 3.30   Clayton Falls BC Hydro's HARP project is an energy storage system
that converts off-peak electricity from a renewable source into hydrogen
via an electrolyzer. The hydrogen is used for energy later on, during times
of peak demand.
(Reproduced from Ref. 42, with kind permission.)

According to the company,[43] renewable hydrogen is a very cost-
effective and convenient way to store renewable energy; it is more economical than importing electricity from Alberta or from the USA to
provide customers with power at peak times.

This insight immediately refers the reader to a sector in which solar
hydrogen will play a role, namely in the production of valued peak
power through the grid, resulting in consistent "peak shaving" of the
price of electricity for the customers of the grid. This phenomenon is
already taking place in Germany and in Italy, where an impressive
13 GW overall of PV power has been installed between 2008 and 2011.
Indeed, thanks to generous feed-in-tariffs and to the concomitant dra-
matic fall in the price of PV modules, between 2008 and 2011 Italy has
become the world's second country for installed PV power, with over
13 GW of grid-connected PV plants present in the early days of 2012.[44]

The impact of all this power on the electricity market started to
become clear in the early months of 2011 (Figure 3.31).[45] The valued PV
power produced in the sunny hours during the day "pushes" out of the
market the most expensive power plants, namely the expensive open
cycle "turbogas" plants running on natural gas only. By doing so, it
effectively cuts the market price of electricity, resulting in a clear benefit
for the grid customers.

Similar findings have been reported from the Hydrogen and Renewables Integration (HARI) project, run from 2003 through late 2011 at
West Beacon Farm in Leicestershire, UK. The project involved the
integration of an electrolyzer, hydrogen storage and fuel cells with an
existing renewable energy system. In full agreement with Bossel's insight

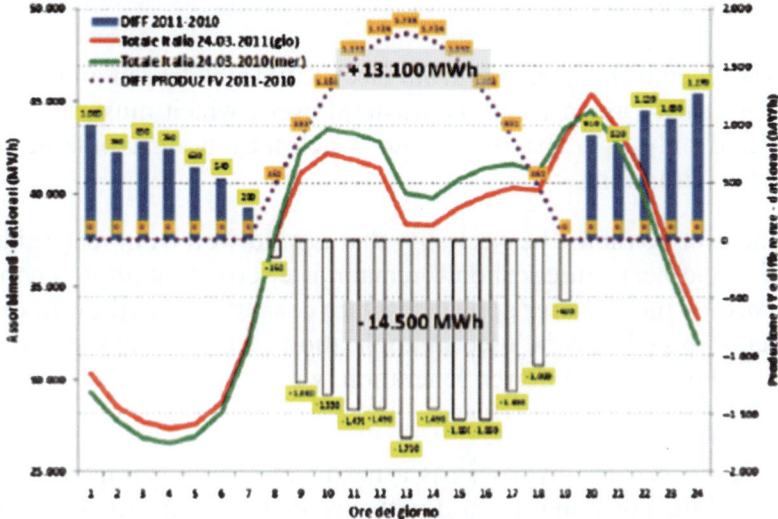

Figure 3.31 Peak shaving in Italy. Hourly load on the transmission grid (green and red curves) and the difference between the same solar days in 2010 and 2011, along with estimated PV generation.
(Reproduced from Ref. 45, with kind permission.)

Figure 3.32 The pressurized hydrogen storage cylinders at West Beacon Farm in Leicestershire, UK, where the HARI project was run from 2003.
(Reproduced from Ref. 46, with kind permission.)

mentioned above, the researchers found that the efficiency of passing through the cycle from electricity to hydrogen and back to electricity is (at typically around 20% or lower) poor (Figure 3.32).[46]

In Gammon's words,

Clearly, converting energy from electricity to hydrogen and back to electricity again is a very wasteful cycle, which must be considered only as a last resort, but which may be unavoidable in certain situations.

Wherever possible, the electricity that is hard-won from renewable (or any other) sources should remain as electricity until it is consumed by the end-user appliance. Once converted to hydrogen, the energy should only be used in applications, such as transport and remote or portable power generation, where only a fuel is able to do the job.

However, because hydrogen is a means of energy storage it can be harnessed for powering the grid with electricity or, directly, for vehicle propulsion and heat generation. Overall, the combined use of fuel cells, electrolyzers and hydrogen makes it possible to use the grid as the major means of energy distribution and supply – for electricity, heat and transportation – thus removing the distinctions between these three energy forms and, at the same time, increasing the overall energy efficiency.

Eventually, as stated by Gammon,[46] complementary solar hydrogen and electricity will be used in a symbiotic partnership to derive maximum benefit in terms of flexibility and efficiency, but electricity that is easy to dispatch will certainly continue to play a large and increasing role in our energy system, even when carbon-neutral solar hydrogen is readily available.

References

1. D. Abbott, *Proc. IEEE*, 2009, **97**, 1.
2. BP, *Statistical Review of World Energy*, 2011. www.bp.com/assets/bp_internet/globalbp/globalbp_uk_english/reports_and_publications/statistical_energy_review_2011/STAGING/local_assets/pdf/statistical_review_of_world_energy_full_report_2011.pdf (last accessed on 26/01/2012). For an interesting overview of this document, see: O.M., "The world gets back to burning", *The Economist*, 8 June 2011, http://www.economist.com/blogs/schumpeter/2011/06/energy-statistics (last accessed on 26/01/2012).
3. K.-A. Adamson (Pike Research), *Fuel Cell Myth #2: There is Not Enough Platinum in the World to Roll Out a Global Fuel Cell LDV*

Fleet, 16 March 2011. http://www.pikeresearch.com/blog/articles/fuel-cell-myth-2-there-is-not-enough-platinum-in-the-world-to-roll-out-a-global-fuel-cell-ldv-fleet

4. U. Bossel, *Proc. IEEE*, 2006, **84**, 1826.
5. L. Zyga, Why a hydrogen economy doesn't make sense, *Physorg. com*, 11 December 2006. www.physorg.com/news85074285.html (last accessed on 30/01/2012).
6. D. Abbott, *Proc. IEEE*, 2010, **98**, 42.
7. A. Steinfeld, *Sol. Energy*, 2005, **78**, 603.
8. US Department of Energy, *Solar Trough Systems*, US Department of Energy Concentrating Solar Power Program, 1998, www. nrel.gov/docs/legosti/fy98/22589.pdf (last accessed on 21/01/2012).
9. D. Humphrey, *Hovering in the wings: Linear Fresnel Technology*, 14 January 2010, http://social.csptoday.com/technology/hovering-wings-linear-fresnel-technology (last accessed on 21/01/2012).
10. H. Knight, The next best thing to oil, *New Scientist*, 12 August 2010, www.newscientist.com/article/dn19308-the-next-best-thing-to-oil.html (last accessed on 21/01/2012).
11. M. Roeb and H. Müller-Steinhagen, *Science*, 2010, **329**, 773.
12. M. Roeb, M. Neises, N. Monnerie, C. Sattler and R. Pitz-Paal, *Ener. Environ. Sci.*, 2011, **4**, 2503.
13. A. G. Konstandopoulos and S. Lorentzou, in *On Solar Hydrogen and Nanotechnology*, ed. L. Vayssieres, John Wiley & Sons, New York, 2010, p. 623.
14. A. W. Weimer, C. Perkins, P. Lichty, H. Funke, J. Zartman, D. Hirsch, C. Bingham, A. Lewandowski, S. Haussener and A. Steinfeld, *Development of a Solar-thermal ZnO/Zn Water-splitting Thermochemical Cycle. Final Report*, 1 April 2009, www.nrel.gov/hydrogen/pdfs/development_solar-thermal_zno.pdf (last accessed on 21/01/2012).
15. *The Hydrosol Projects*, www.hydrosol-project.org (last accessed on 21/01/2012).
16. M. Roeb, J.-P. Säck, P. Rietbrock, C. Prahl, H. Schreiber, M. Neises, D. Graf, M. Ebert, W. Reinalter, M. Meyer-Grünefeldt, C. Sattler, A. Lopez, A. Vidal, A. Elsberg, P. Stobbe, D. Jones, A. Steele, S. Lorentzou, C. Pagkoura, A. Zygogianni, C. Agrafiotis and A. G. Konstandopoulos, *Sol. Energy*, 2011, **85**, 634.
17. A. G. Konstandopoulos, Solar hydrogen from thermochemical water-splitting: the hydrosol process and beyond, in *The Fuel Cells and Hydrogen Joint Technology Initiative*, Stakeholders General Assembly 2009, 26–27 October, Brussels. http://ec.europa.eu/

research/fch/pdf/konstandopoulos.pdf#view=fit&pagemode=none (last accessed on 21/01/2012).

18. C. Agrafiotis, M. Roeb, A. G. Konstandopoulos, L. Nalbandian, V. T. Zaspalis, C. Sattler, P. Stobbe and A. M. Steele, *Sol. Energy*, 2005, **79**, 409.

19. O. Deutchmann and A. G. Konstandopoulos, in *Handbook of Combustion*, ed. M. Lackner, F. Winter and A. K. Agarwal, Wiley VCH, Weinheim, 2010, vol. 2, p. 465.

20. European Commission, *Hydrogen: vector for clean energy*, 15 June 2007, http://ec.europa.eu/research/star/index_en.cfm?p=22_main (last accessed on 21/01/2012).

21. T. Kodama, N. Gokon and T. Yamamoto, *Sol. Energy*, 2008, **82**, 73.

22. *Hydrosol-3D Project*, www.hydrosol3d.org (last accessed on 21/01/2012).

23. A. G. Konstandopoulos, Solar reactors for hydrogen and carbon neutral fuel production: Hydrosol technology and beyond, *International Conference on Hydrogen Production (ICH2P-2010)*, Istanbul, 16–18 June 2010. The research is currently being carried out in the framework of the project "Advanced Multifunctional Reactors for Green Mobility and Solar Fuels" (ARMOS) funded by an Advanced Grant from the European Research Council.

24. A. Zygogianni, C. Pagoura, G. Karagiannakis, C. Agrafiotis and A. G. Konstandopoulos, Solar carbon-neutral syngas generation by H_2O and CO_2 splitting thermochemical cycles, *International Conference on Hydrogen Production (ICH2P-2011)*, Thessaloniki, 19–22 June 2011.

25. W. C. Chueh, C. Falter, M. Abbott, D. Scipio, P. Furler, S. M. Haile and A. Steinfeld, *Science*, 2010, **330**, 1797.

26. L. Campenni, Solar fuel: no more drilling!, *Optics & Photonics Focus*, 11 May 2011. http://opfocus.org/index.php?topic=story &v=13&s=4 (last accessed on 30/01/2012).

27. S. C. Roy, O. K. Varghese, M. Paulose and C. A. Grimes, *ACS Nano*, 2010, **4**, 1259.

28. W. Chueh and S. Haile, *Philos. Trans. R. Soc. London, Series A*, 2010, **368**, 3269.

29. P. Furler, J. R. Scheffe and A. Steinfeld, *Energy Environ. Sci.*, 2012, DOI: 10.1039/C1EE02620H.

30. G. B. Haxel, J. B. Hedrick and G. J. Orris, Rare earth elements – Critical resources for high technology, *US Geological Survey Fact Sheet*, 087-02, Reston, VA, 2002.

31. W. C. Chueh and S. M. Haile, *Philos. Trans. R. Soc. London, Series A*, 2010, **368**, 3269.

32. M. Allen, *Solar Fuel Invention Seeks Industrial Partner*, www. swissinfo.ch/eng/science_technology/Solar_fuel_invention_seeks_ industrial_partner.html?cid=29182862 (last accessed 06/01/ 2011).

33. R. Kunzig and W. Broecker, *New Scientist*, 12 January 2009, 2690.

34. J. E. Miller, M. D. Allendorf, R. B. Diver, L. R. Evans, N. P. Siegel and J. N. Stuecker, *J. Mater. Sci.*, 2008, **43**, 4714.

35. M. Lavelle, *National Geographic News*, 10 August 2011, http:// news.nationalgeographic.com/news/energy/2011/08/110811-turning-carbon-emissions-into-fuel (last accessed on 30/01/2012).

36. R. B. Diver, J. E. Miller and N. P. Siegel, *ASME 2010 4th International Conference on Energy Sustainability, Volume 2, Solar Thermochemistry*, Paper no. ES2010-90093 pp. 97–104. http:// dx.doi.org/10.1115/ES2010-90093 (last accessed on 30/01/2012).

37. N. S. Lewis and D. G. Nocera, *Proc. Natl. Acad. Sci. U.S.A.*, 2006, **103**, 15729.

38. C. Day and K. Gibbons, *The Grid*, The Center for Culture, History, and Environment at the University of Wisconsin-Madison, http:// envhist.wisc.edu/cool_stuff/energy/grid.shtml#_ftnref1 (last accessed on 30/01/2012).

39. US Department of Energy, *What is the Smart Grid?* www.smartgrid. gov/the_smart_grid (last accessed on 30/01/2012).

40. *Solar + Hydro power: winning combination?* 3 November 2010, www.smartgridelectronics.com/2010/11/solar-hydro-power-winning-combination.html (last accessed on 30/01/2012).

41. www.bchydro.com/community/recreation_areas/clayton_falls_ recreation_site.html (last accessed on 30/01/2012).

42. Powertech Inc., *Hydrogen Assisted Renewable Power System*, 4 February 2011, www.powertechlabs.com/temp/20112439066/ HARP_DataSheet_Feb_4_2011web.pdf (last accessed on 30/01/ 2012).

43. *Hydropower to hydrogen: BC Hydro testing a new approach to large-scale energy storage*, 12 November 2010, www.smartgridnews.com/ artman/publish/Technologies_Storage/Hydropower-to-hydrogen-A-new-approach-to-energy-storage-3270.html (last accessed on 30/01/2012).

44. For updated statistics on PV power in Italy, see the URL: http:// atlasole.gse.it/atlasole/ (last accessed on 30/01/2012).

45. F. Meneguzzo, HELIONOMICS – Utility-scale PV plants in Italy: A sustainable way towards low price electricity? *Sun New Energy Conference – SuNEC 2011*, Santa Flavia, Italy, 7 July 2011. www.

slideshare.net/fmeneguzzo/fmeneguzzo-sunec-2011-palermo-italy
(last accessed on 30/01/2012).

46. R. Gammon, A. Roy, J. Barton and M. Little, *Hydrogen Renewables and Integration (HARI)*, case study report for the International Energy Agency, March 2006, http://ieahia.org/pdfs/
HARI.pdf (last accessed on 30/01/2012).

CHAPTER 4
Solar Hydrogen Utilization

4.1 Hydrogen Fuel Cell Engines

Invented in 1839, the hydrogen fuel cell has been employed widely since the early 1960s in space probes to generate onboard electricity, water and heat.

Indeed, in the fuel cell the controlled reaction of hydrogen with oxygen yields electricity, heat, and water, *directly* converting into electrical energy the chemical energy of the bound H_2 molecule.

In most fuel cells developed thus far the H_2 dissociation reaction is usually catalyzed by platinum at the anode's surface and takes place at a temperature of approximately $80\,°C$ (Figure 4.1).

The presence of easily poisoned platinum requires the use of high purity "technical-grade" (purity as high as 99.999%) H_2 because the carbon- and sulfur-containing impurities of "commercial-grade" hydrogen would quickly degrade, reducing the life of the fuel cell stack (Figure 4.2).

Indeed, similar to what happens with solar cells, an individual fuel cell (about 2 mm thick) generates a comparatively low potential of less than 1 V. Hence, for practical application several hundred cells are connected in series to form a so-called "stack". A 200 V system potential, for example, is used to power a fuel cell vehicle (FCV) such as the Mercedes *B-Class F-CELL*, the first fuel cell passenger car to be produced under series conditions, which attains an operating range of around 400 km with its 700 bar hydrogen tank in the sandwich floor unit.[1]

From a technical viewpoint only, the clean, efficient and compact H_2-fueled fuel cell fits excellently into the ongoing electrification trend;[2] it

Solar Hydrogen: Fuel of the Future
Mario Pagliaro and Athanasios G. Konstandopoulos
© Mario Pagliaro and Athanasios G. Konstandopoulos 2012
Published by the Royal Society of Chemistry, www.rsc.org

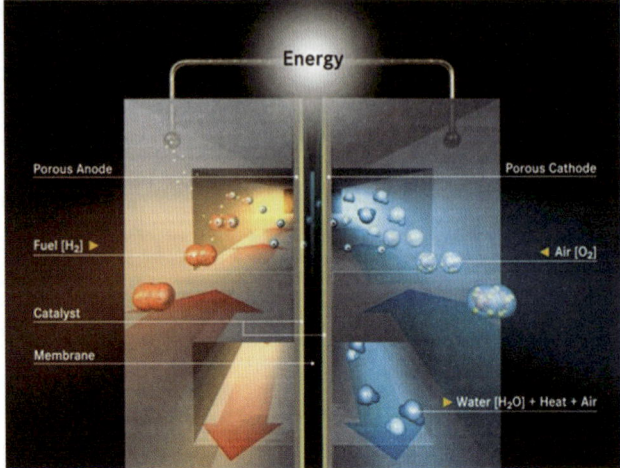

Figure 4.1   A PEM fuel cell normally uses a Pt catalyst at the anode's surface and a proton exchange membrane, usually made of Nafion, an expensive fluorinated polymer functionalized with sulfonic functional groups. (Reproduced from Ref. 1, with kind permission.)

Figure 4.2   The fuel cell stack used in the Mercedes-Benz B-Class F-CELL. Manufacture of a small series of this car commenced in late 2009. (Reproduced from Ref. 1, with kind permission.)

serves as a battery recharging device for long-life power packages in portable electronics, as stationary combined heat and power (CHP) facilities, and as an electrical generator to replace the current poorly efficient mobile electrical generators (lead batteries and diesel gensets).[3]

Moreover, the H_2-fueled fuel cell is without moving parts and, thus, is vibration-free and noiseless; it is no heavier than the internal combustion engine (ICE), and fits into a conventional engine compartment without major modifications. Overall, the combination of the fuel cell *and* an electric motor is 2–3 times more efficient than an ICE.[4]

The installed cost of a hydrogen fuel cell system depends on the technology, configuration and size. In general, installation costs of a fuel cell system range from $5000 kW^{-1} to $10 000 kW^{-1},[5] meaning that the fuel cell is technically, but not *economically*, more efficient than an internal combustion engine.

Currently, H_2 fuel cells are mostly present in German, Greek and Italian submersibles of the respective navies. The main reason for this market failure is due to their inherently high costs. Legacy fuel cell technologies such as proton exchange membranes (PEMs), phosphoric acid fuel cells (PAFCs), and molten carbonate fuel cells (MCFCs) have all required expensive precious metals, corrosive acids, or hard to contain molten materials. From an environmental viewpoint, hydrogen-powered fuel cell vehicles achieve reductions in greenhouse gas emission below 1990 levels by 80% or more (hydrogen ICE hydrogen powered vehicles by 60%) with near elimination of urban air pollution.[6] The car maker Honda, for example, in 2007 unveiled the world's first fuel cell vehicle (Figure 4.3), an electric car powered by H_2 but only offering a driving range of 240 miles, which is currently available to lease in the USA.

After all, on average a vehicle exhaust catalyst (Figure 4.4) contains around 1–3 g of platinum group metals (PGM; about 1.5 g platinum, 0.5 g palladium and 0.1 g rhodium).[7] Perhaps it would be more logical to

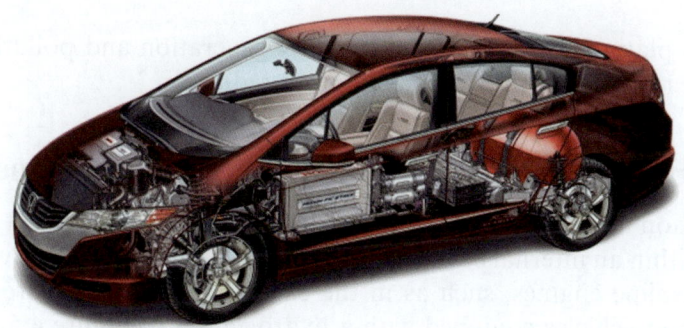

Figure 4.3 In 2007 Honda's FCX Clarity was the world's first fuel cell vehicle to be commercialized.
(Reproduced from www.manufacturingdigital.com, with kind permission.)

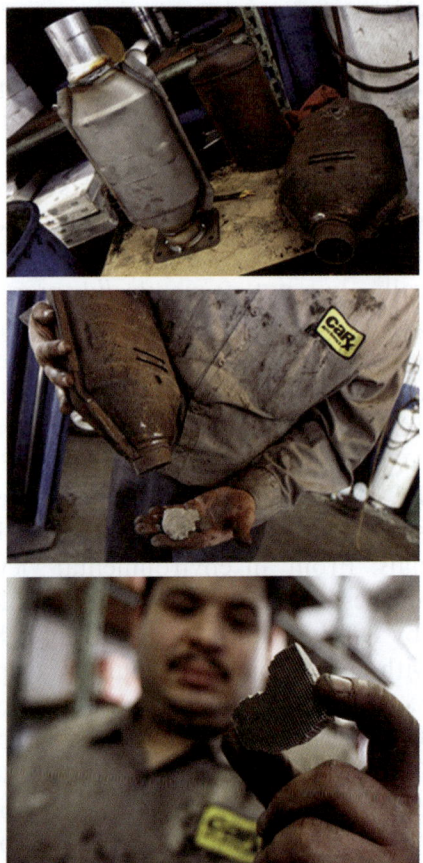

Figure 4.4   A catalytic converter and the honeycomb filter containing platinum.
(Image of Joshua Lott.
Reproduced from Ref. 37, with kind permission)

use this platinum in fuel cells for energy generation and pollution *prevention*, rather than in cars for pollution reduction.[8]

4.2 Hydrogen Fueled Internal Combustion Engines

In addition to its indirect use in fuel cells, hydrogen can be burnt directly in air within an internal combustion engine, with remarkable advantages over gasoline engines, such as in the case of the BMW Hydrogen 5th generation vehicles equipped with a hydrogen tank (Figure 4.5).[9]

Curious as it may seem, the reciprocating ICE operated in the Otto- or Diesel mode that came to market in the late 19th century is still the dominating power-train technology in 2012. However, an Otto cycle

Figure 4.5 The Hydrogen BMW 5 was unveiled at the Expo 2000 Exhibition. (Reproduced from Ref. 7, with kind permission.)

internal combustion engine running on hydrogen has a maximum efficiency of about 38%, 8% higher than the gasoline ICE. In addition, hydrogen engine conversion technology is more economical than fuel cells,[10] and the ICE is familiar to engineers and craftsmen in the car industry as well as in repair shops. In general, the reciprocating *hydrogen internal combustion DI engine* has high power density with regard to volume and weight, is highly efficient and nearly emission free.

The idea is that, in order to be successful on the market, the characteristics of daily operation and performance of a hydrogen vehicle should be comparable to those of a conventional gasoline or diesel vehicle. After 25 years of research in the field, which started during the 1970s oil crisis, BMW is aware that this requirement can only be met by the installation of a hydrogen ICE (Figure 4.6) which uses energy dense liquid hydrogen (Figure 4.7).

The properties of gaseous hydrogen are significantly different from those of gasoline (Table 4.1). Liquefied hydrogen has lower energy density by volume than gasoline, by approximately a factor of 4, because of the low density of liquid hydrogen. Obviously, gaseous hydrogen has a very low density, which entails a lower density of the air–fuel mixture, ρ_G, and thus an 18% lower mixture calorific value, H_G, in the external mixture hydrogen mode.

However, in the *internal* mixture mode (Figure 4.8) the low density of hydrogen is *not* relevant because the pressurized hydrogen is fed to the cylinder by a direct injection system that affords a 17% higher mixture calorific value for hydrogen when compared with gasoline.

Table 4.1 also shows a typical property of hydrogen, namely its capability for ignition within a wide range of air : hydrogen ratios, which

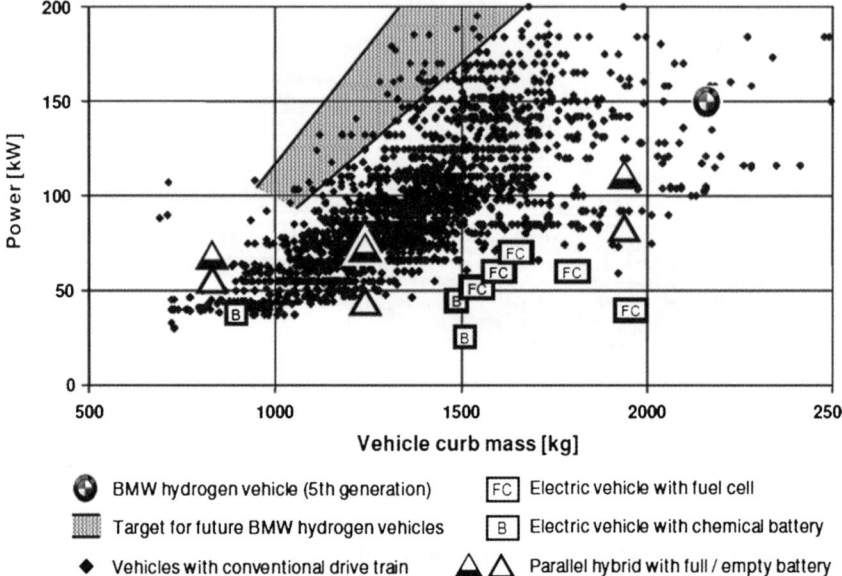

Figure 4.6 Power/mass characteristics of vehicles with conventional and alternative
drive train systems.
(Reproduced from Ref. 7, with kind permission.)

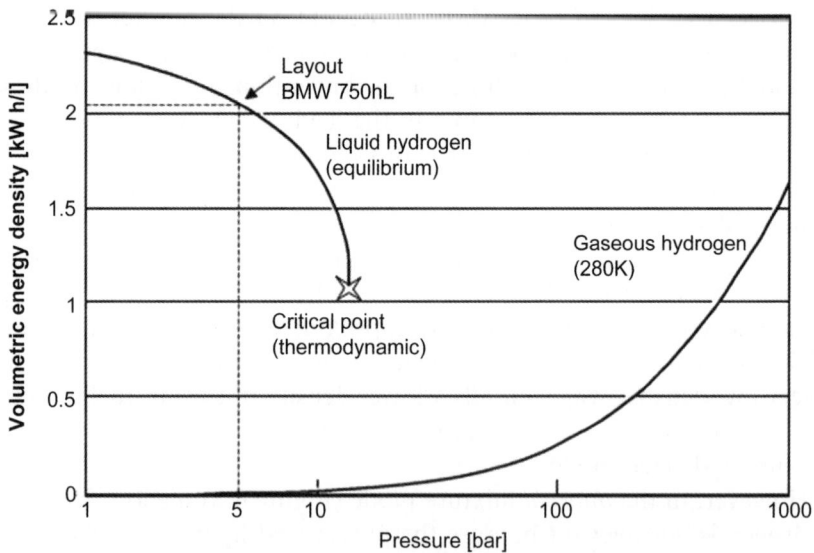

Figure 4.7 Power/mass characteristics of vehicles with conventional and alternative
drive train systems. At room temperature and pressure, the density of
hydrogen is so low that it contains less than 1 : 300 the energy in an equivalent
volume of gasoline.
(Reproduced from Ref. 7, with kind permission.)

Table 4.1 Physical properties of hydrogen and gasoline. (Adapted from Ref. 9, with kind permission.)

Property	Hydrogen	Gasoline
Density (ρ)	0.09 kg m^{-3}	730 780 kg m^{-3}
Ignition limits in air	4–76 (vol%)	1–7.6 (vol%)
Flame velocity (at $\lambda = 1$)	2.0 m s^{-1}	0.4–0.8 m s^{-1}
Density of stoichiometric mixture (ρ_G)	0.94 kg m^{-3}	1.42 kg m^{-3}
Mixture calorific value,[†] H_G	3.2 MJ m^{-3}	3.9 MJ m^{-3}
Mixture calorific value,[‡] H_G	4.5 MJ m^{-3}	3.8 MJ m^{-3}

[†]Port injection.
[‡]Direct injection.

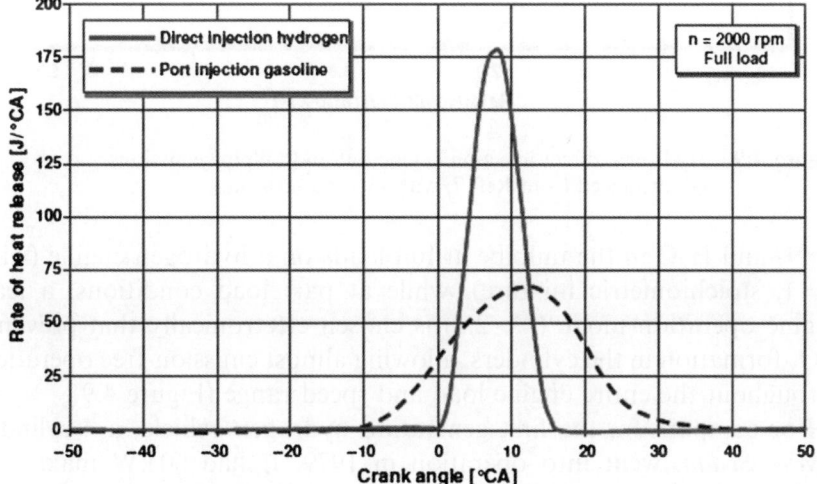

Figure 4.8 Full-load combustion behavior of direct injection hydrogen and port injection gasoline engines.
(Reproduced from Ref. 7, with kind permission.)

allows a remarkable increase in efficiency because of leaner combustion. The high flame propagation velocity of air–hydrogen mixtures entails outstanding combustion properties with significantly shorter combustion periods in the full-load range compared with gasoline engines.

Overall, owing to the ability to realize ideal combustion control with high compression ratios, scientists at BMW aim to increase the effective efficiency of a hydrogen DI internal combustion engine to 50%, compared with the current 37% (Figure 4.8).

Finally, prolonged experimentation has shown that NO$_x$ tailpipe emissions can be minimized to a few ppm by concomitant employment of a simple reduction catalyst for converting the NO$_x$ and unburned H$_2$

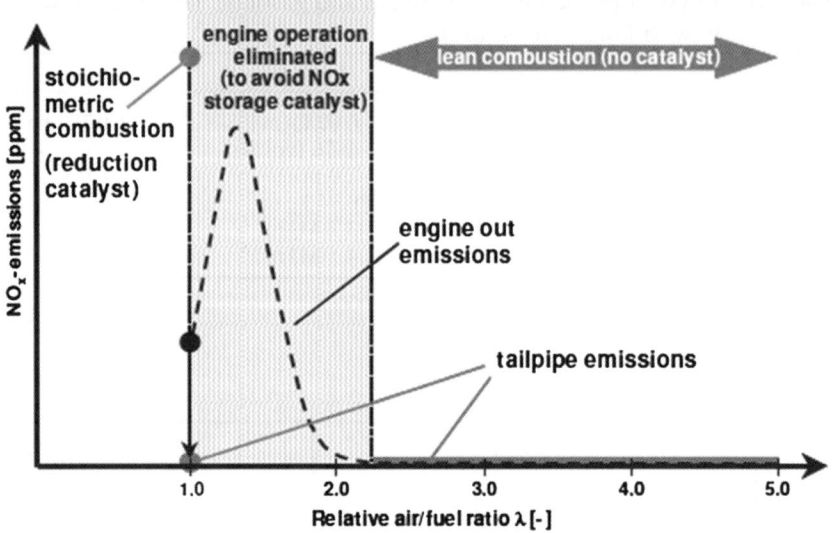

Figure 4.9 Hydrogen direct injection engine out and tailpipe emissions.
(Reproduced from Ref. 7, with kind permission.)

to N_2 and H_2O in the tailpipe at full loads on a hydrogen engine (with $\lambda = 1$, stoichiometric mixture), while at part load conditions, a lean engine operation mode ($\lambda > 2.2$) is chosen electronically that prevents NO_x formation in the cylinders, allowing almost emission-free operation throughout the entire engine load and speed range (Figure 4.9).

For comparison, the first generation hydrogen vehicle, a 4-cylinder BMW *Sedan*, went into operation in 1979. It had 60 kW maximum power and a top speed of 160 km h^{-1} with a range of operation around 400 km. The 5th generation vehicles, launched in 2000, had 150 kW maximum power and could be operated in a dual mode (hydrogen and gasoline), allowing them to extend their range of operation from about 300 km in the hydrogen mode to a total of about 900 km. For further development, future hydrogen ICEs will have higher exergy efficiencies as a result of reducing inherent irreversibilities, through utilizing the huge amounts of waste heat in the cooling system and the exhaust.

4.3   Hydrogen Motoring: A Dream Never Coming True?

With about 600 million passenger vehicles around the globe today – and currently with a worldwide production of approximately 55 million units per year – the automobile powered by fossil oil-based fuels is a major source of environmental pollution. Efforts to market hydrogen-powered cars, which started in the early 2000s, have generally failed. In 2010, *The Economist* concluded that:

> Having soared on the promise of carbon-free motoring, the idea of the 'hydrogen economy' crashed and burned when it collided with reality.

> Hundreds of experimental hydrogen-powered cars – once hailed as the best solution for reducing America's dependence on foreign oil for over half its consumption – are now gathering dust in manufacturers' parking lots.[11]

In 2009, the US Government cancelled funding for research into hydrogen-powered vehicles. In the same country, General Motors ditched its fleet of 100 Chevrolet *Equinox* fuel-cell cars after a two-year trial. Likewise, in mid 2009 BMW withdrew its own test fleet of 100 *7-Series* limousines equipped with hydrogen internal combustion engines, tested since 2000.[12]

However, believing that the combination of a fuel cell electric vehicle and the solar hydrogen station could help lead to the establishment of a hydrogen society based on renewable energy, the Japanese car maker Honda is continuing to build a network of refueling stations to accommodate a growing customer base both in Europe (Figure 4.10)

Figure 4.10 The first hydrogen station in the UK, at the Honda factory at South Marston, located near the car maker's plant in Swindon, Wiltshire, has been operated by industrial gas company BOC since September 2011. (Reproduced from Ref. 11, with kind permission.)

and in the USA (see discussion of the new generation of solar hydrogen stations for home use, based on the high differential pressure electrolyzer, Chapter 2).[13]

Similarly, Daimler's Mercedes-Benz also recently unveiled that they would launch a mass-produced fuel cell car in 2014, based on the next-generation *B-Class* model. In 2010, Toyota expanded its hydrogen car program, aiming to come to market in 2015 or earlier with a vehicle that will be reliable and durable, with excellent fuel economy and zero emissions at an affordable price The carmaker has cut the cost to make hydrogen models for less than $100 000 and aims to halve that price by the time sales begin.[14]

A simple look at the comprehensive and continuously updated overview of all types of hydrogen vehicles (since, astonishingly enough, 1807) offered by the *H2mobility.org*[15] website clearly shows that hydrogen motoring is far from being a doomed reality, as the author of the cited article in *The Economist* seems to believe.

In North America, Japan and Europe, demonstration fleets of hydrogen-fueled fuel cell or ICE buses are on the road, operated by municipal transport professionals and run under conventional conditions with passenger loads, and these maintain the usual time schedules.

The experience is positive. For example, 36 Mercedes-Benz *Citaros*, fitted with hybrid fuel cell and battery technology and deployed on the streets on three continents from 2003 to 2009 have logged more than 2.2 million kilometers by late August 2011. No hazard has been reported, and no major repairs were necessary (Figure 4.11).

Figure 4.11   The Citaro Fuel Cell bus made by Daimler runs on hydrogen only. The bus carries seven cylinders on the roof, containing 35 kg of hydrogen in all. (Reproduced from http://green.autoblog.com, with kind permission.)

Given that most city transit buses run on the same routes, only a small number (possibly as low as one) of hydrogen fueling stations are needed to supply the fuel. As the hydrogen fueling infrastructure continues to expand so will bus routes, including for buses that travel to cities farther away and need to refuel at a second station.

Clearly, the development of a public hydrogen supply infrastructure is essential for the successful marketing of both hydrogen fuel cell and ICE vehicles. What remains to be done, therefore, is to build a nationwide infrastructure of hydrogen refueling stations, such as in Germany, where the first filling stations have already become established in metropolitan areas such as Berlin und Hamburg. The more a station is used the lower the cost of the hydrogen supplied, with an almost hyperbolic cost trend (Figure 4.12).

Currently, 7 of the 30 hydrogen filling stations in Germany are integrated into a public filling station operation, whereas Daimler and the Linde agreed in 2009 to build 20 additional hydrogen filling stations.

Eventually, the whole structure of the automobile and trucking industry, with the ICE at its core, will be replaced by an industrial structure built around an electrochemical motor requiring the production of membranes and stacks, heat exchangers, hydrogen tanks and delivery systems, electrical and electronic equipment.

Electric cars that use either H_2 fuel or electricity from batteries will certainly be competitive because battery electric vehicles (BEVs) are typically 3 to 4 times as efficient as hydrogen powered vehicles (Figure 4.13), and BEVs can be recharged easily at home every two or

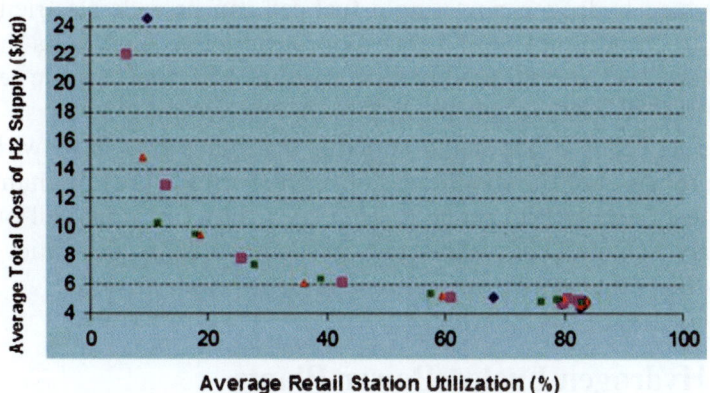

Figure 4.12   Hydrogen fuel supply cost as function of filling station utilization. (Reproduced from Ref. 18, with kind permission.)

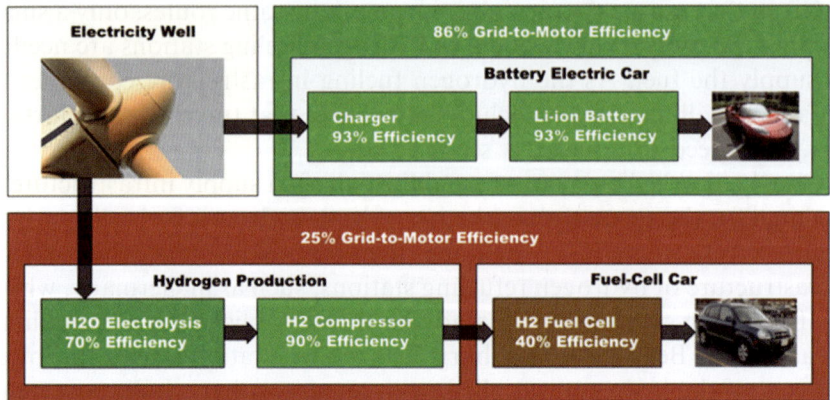

Figure 4.13   The infrastructure for the electric car exists already everywhere: the electric grid. Therefore the high investment costs of a hydrogen station need not be incurred.
(Reproduced from Ref. 14, with kind permission.)

three days (with no need for an extensive infrastructure) using the *existing* grid and plugs.

The well-to-wheels efficiency of hydrogen FCVs tends to be about one-third that of electric vehicles (EVs) when electrolysis is used.[16] In contrast, BEVs, which are already commercially available from several manufacturers, do not require expansion of the existing infrastructure for electricity transmission and distribution, and ideally can be recharged at night with idle off-peak power plant capacity that currently goes unused.[17]

However, compared with BEVs, automobiles equipped with hydrogen fuel cells generate electricity *onboard* and have a far longer autonomy. Compressed hydrogen powering a fuel cell can provide electricity to a vehicle traction motor with 5 times more energy per unit mass than the current NiMH batteries used in most gasoline hydrogen electric vehicles, and 2 times more than advanced lithium-ion batteries.[18]

The fuel cell EV has many superior attributes over EVs with only 320 km ranges, but the advantages of the fuel cell EV are dominant if the BEV must have 480 km range to serve as a fully functional all-purpose passenger vehicle. This is plotted in Figure 4.14 as the ratio of the battery EV value to the fuel cell EV value for each attribute.

4.4   Hydrogen Fueled Power Plants

Hydrogen can be used efficiently as fuel for thermoelectric power plants. In 2010, Italy's largest electricity utility (Enel) opened the world's

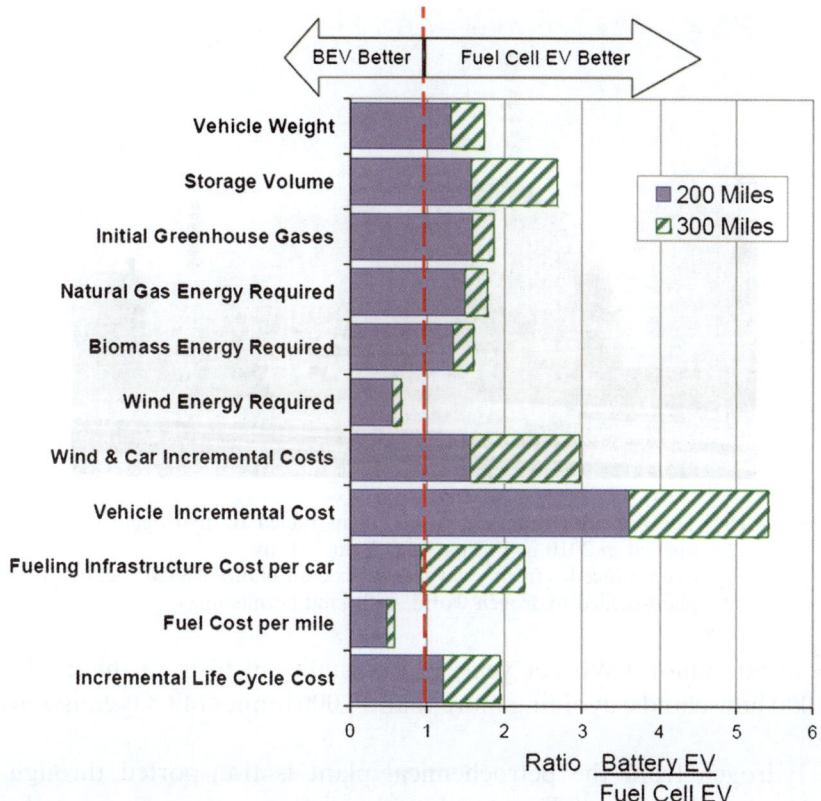

Figure 4.14 The useful specific energy (energy per unit mass) of hydrogen and fuel cell systems (including a peak power battery), compared with the useful specific energy of various battery systems.
(Reproduced from Ref. 16, with kind permission.)

first hydrogen power combined cycle plant near Venice (Figure 4.15). Hydrogen is supplied using specially built pipelines from the nearby Polimeri Europa petrochemical plant, where ethylene-cracking is carried out.

The 12 MW power plant comprises a hydrogen-fueled combined cycle plant and burns hydrogen gas in a turbine capable of resisting hydrogen embrittlement, which was developed in partnership with General Electric and generates both electricity and heat. The plant uses 1.3 tonnes of hydrogen per hour, has an overall efficiency of about 42%, and is essentially free of emissions. The efficiency of the process is increased by using the heat from the emissions to generate high-temperature steam, which is sent to the nearby coal-fired plant to generate an additional 4 MW of power capacity. Overall, the electricity generated, equal to

Figure 4.15   The first industrial-size power plant fueled by hydrogen in the world
opened in 2010 in Fusina, near Venice, Italy.
(Reproduced from www.demotix.com/photo/388950/enel-first-power-
plant-fuelled-hydrogen-world, with kind permission.)

about 60 million kWh per year, will be sufficient to meet the needs of
20 000 households, avoiding more than 17 000 tonnes of CO_2 emissions a
year.

Hydrogen from the petrochemical plant is transported through a
pipeline made of specially coated steel, which resists hydrogen embrittlement (Figure 4.16).

Hydrogen gas piping is routine in large oil refineries, because
hydrogen is used to hydrocrack fuels from crude oil. As a result, millions
of tonnes of elemental hydrogen are distributed around the world
through these pipelines each year. For example, the first long (240 km)
hydrogen pipeline was built by Linde Germany, in the Ruhr area, in
1938. The pipeline is still in operation nowadays, connecting some 14
production sites with an estimated capacity of 250 million Nm^3 of
hydrogen per year.[19] The longest hydrogen pipeline in the world is a
250 mile line between Belgium and France. Similarly, about 700 miles of
hydrogen pipelines now operate in the USA, generally near oil refineries.

Once built, hydrogen pipelines are the cheapest and most effective way
to distribute large volumes of hydrogen. Contrary to common opinion
surrounding the so-called "hydrogen infrastructure", hydrogen distribution through pipelines is *not* problematic, just more expensive than,
for example, the distribution of natural gas.

Building the infrastructure in a systematic way is largely affordable.
For example, a study[20] by researchers from General Motors and Shell

Figure 4.16 The hydrogen pipeline carrying the fuel from the petrochemical station to the power plant in Fusina (Venice), Italy.

Hydrogen clearly showed that an investment of US$10–15 billion, comparable to one-half the cost of the Alaskan pipeline, would be sufficient to establish 12 000 hydrogen stations in the USA, putting hydrogen within 2 miles of 70% of the US population.

Similar to the Italian Enel, the Germany company Enertrag, together with Total, Vattenfall and Deutsche Bahn, built in 2010, in a lightly populated area east of Berlin, a hydrogen-hybrid power plant that combines wind energy sources to deliver electricity to 2000 homes and heat to 300 households with no carbon emissions (Figure 4.17).[21]

The Uckermark operation is a hybrid plant that relies on three wind turbines and water electrolysis for energy storage in hydrogen fuel. Hydrogen tanks store the extra power generated by the wind turbines so that the plant's stored fuel can be used to run hydrogen-powered vehicles or can be combined with biogas to produce electricity during low-wind conditions. In addition, the renewable hydrogen is used in hydrogen refueling stations located in Berlin and Hamburg.

The storage of large quantities of hydrogen functions therefore as grid energy storage, an essential requirement for the forthcoming hydrogen energy infrastructure. As in the case of natural gas, which is currently stored in large amounts in underground caverns, salt domes and depleted oil and gas fields, hydrogen will be stored in underground caverns, as done for many years by ICI with no difficulties; similarly,

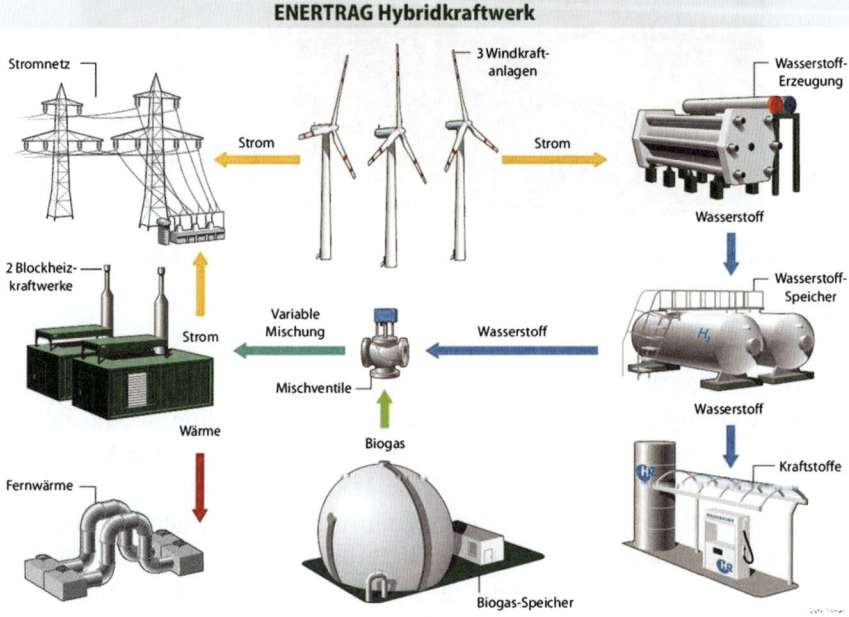

Figure 4.17   Hydrogen-Hybrid Power Station in Prenzlau. This hydrogen power plant meets the needs of local citizens at a cost of 21 million Euros, or €10 500 per customer.
(Reproduced from Ref. 19, with kind permission.)

ConocoPhillips at its Clemens Terminal in Texas has stored hydrogen since the 1980s in a solution-mined salt cavern.[22]

4.5   Hydrogen Energy for Distributed Generation

Hydrogen will soon be extensively applied in distributed systems of energy production that use new, reliable and sustainable hydrogen fuel cells to produce clean energy 24 hours per day, 365 days per year to meet the demanding needs of apartment blocks, office buildings, stores and neighborhoods. Recently, the US company *Bloom Energy* has installed

Figure 4.18 A Bloom Energy "Energy Server" uses SOFCs to produce enough power to account for the entire energy demand of a typical office building. (Reproduced from Ref. 21, with kind permission.)

several newly developed solid oxide fuel cells (SOFCs) at a dozen large US firms that now self-generate electrical power from natural gas (Figure 4.18).[23]

Bloom's fuel cells can flexibly use any fuel, including bio- or natural gas, ethanol and, of course, hydrogen. Thanks to a 30% federal tax credit, they produce silent, low-emission power for less than c$10 per kWh (ten cents per kilowatt-hour), much the same as a combined-cycle gas-turbine plant, but without the noise and fumes.[24]

The so called "Energy server", the self-contained generating unit, costs around $750 000 per 100 kW block, and, like other SOFCs, makes use of a common sand-like powder instead of precious metals like platinum or corrosive materials like acids. It operates at a high temperature (typically above 800 °C) which gives it extremely high electrical efficiency and fuel flexibility, both of which contribute to better economics. However, it also creates engineering challenges that were solved by the company with prolonged financial back-up ($400 million) of venture capital, which started in 2002. The result of this R&D effort was a 25 W fuel cell (Figure 4.19) that makes distributed generation so reliable that the company offers customers the opportunity to purchase the service (clean electricity) for 10 years in place of the product (the Bloom Box), thus avoiding the initial investment and giving immediate cost savings.

Figure 4.19   The fuel cell developed by Bloom Energy is made of thin white ceramic plates (100 × 100 mm). Originally introduced in the late 1990s by K. R. Sridhar and his team at the University of Arizona as part of the NASA Mars space program, these SOFC were later developed by Bloom Energy. (Reproduced from Ref. 21, with kind permission.)

Each ceramic plate is coated with a green nickel oxide-based ink on one side (anode) and another black ink (probably Lanthanum strontium manganite) on the other side (cathode). The electrolyte materials may include a scandia-stabilized zirconia (with higher conductivity than yttria-stabilized zirconia at lower temperatures), which provides greater efficiency and higher reliability when used as an electrolyte in SOFC applications.[25]

Adding another revealing clue to the silent and yet powerful shift occurring in industry, from polluting technology to cleantech, in 2011 Bloom Energy purchased from the automaker Chrysler a former car manufacturing factory in Newark, Delaware, which was closed down during the 2008 recession. The factory is currently being repurposed and will eventually manufacture 30 MW modules using its SOFCs.

Innovation in fuel cell technology and solar hydrogen is rapidly occurring in several countries. In Italy, Electro Power Systems has developed a self-recharging back-up power system (*ElectroSelf*, Figure 4.20) that overcomes one of the biggest obstacles to mass deployment of fuel cells, namely the sourcing of hydrogen.[26]

The system uses an electrolyzer and a PEM (proton exchange membrane) fuel cell to minimize the mismatch in energy production and consumption by efficiently storing energy from the grid or when renewables are plentiful, and instantaneously releasing energy whenever

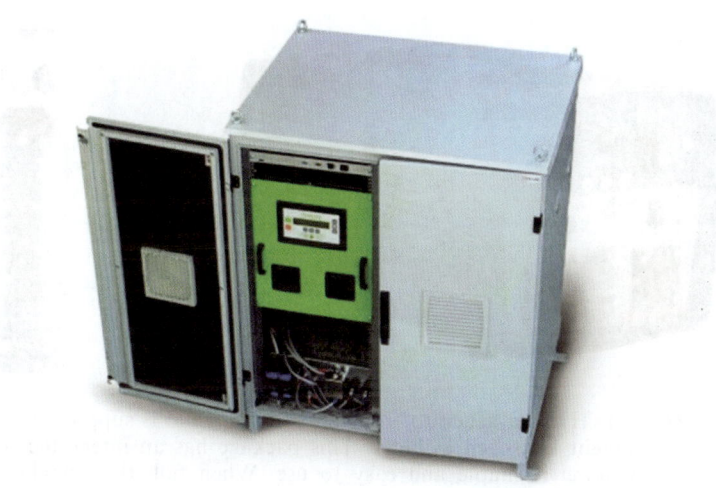

Figure 4.20   By simply adding water to the ElectroSelf, the system self-generates hydrogen fuel using electric power from the grid or from PV solar or wind energy. It releases energy instantaneously when there is a power dip or outage.
(Reproduced from Ref. 24, with kind permission.)

renewables are weak or absent. As such, the system is able to ensure a steady power supply for virtually any application, ensuring always-on power and connectivity. By late 2011, over 600 *ElectroSelf* fuel cell power systems had been installed in Europe, Asia, Africa and the Americas.

To get an idea of the environmental and economic relevance of this fuel-cell technology, it is sufficient to remind the reader that currently there are about 5 million telephone communication towers in remote, off-grid locations that, continuously needing power, rely on either diesel generators or solar PV panels coupled to lead batteries. The *Electro-Self* – which was redesigned in 2011 to be smaller and more efficient, with higher performance and enhanced flexibility – is logistics-free and drastically reduces the economic and environmental economics of back-up power provision. Customers can get rid of batteries and diesel generators, and thus eliminate the costs of fuel logistics and heavy maintenance, as well as the fossil fuel price uncertainty.

Similary, BOC has recently introduced to the portable power market its *Hymera* hydrogen fuel cell generator,[27] which can provide up to 200 W of off-grid DC power (Figure 4.21). When coupled to a portable hydrogen cylinder, the Hymera DC can provide 2–3 kWh of energy. Cylinders can be manifolded together for much longer run-times. Again, water vapor is the only by-product and operation is near-silent, making

Figure 4.21   The Hymera generator runs on hydrogen and is supported by a light-
weight hydrogen cylinder. This package has an integrated regulator,
making it simple and easy to use. When full, the cylinder package
weighs approximately 10 kg.
(Reproduced from Ref. 27, with kind permission.)

these generators ideally suited for applications where the emissions from
diesel or petrol generators could become a problem, or where noise is an
issue. Applications include portable power applications and those that
need to operate continuously for long periods of time, such as security
cameras, environmental monitoring, wireless communication systems
and back-up power for communications systems.

4.6   Portable Devices Running on Hydrogen

Two billion chargers are sold each year through the mobile phone
charger market. The travel charger sub-segment is the fastest growing charger segment, with an estimated market value of more
than €11 billion. A number of new fuel cell companies using
hydrogen fueled fuel cells are now actively trying to establish sales in
this market.

Singapore-based Horizon Fuel Cell Technologies[28] is a fuel cell
manufacturer that offers a wide range (10 W to 5 kW) of standard PEM
fuel cell systems, as well as customized fuel cell system configurations up
to 30 kW.

The company recently commercialized the table-top hydrogen
refueling station *Hydrofill* (Figure 4.22) and the portable emergency fuel
cell back-up generator *HydroPak*. The *Hydrofill* system allows consumers to refill solid state cartridges in a simple way, using water and
electricity as the only inputs.

Figure 4.22   The Hydrofill system is designed to refill Horizon's Hydrostik solid hydrogen cartridges using a high performance water electrolyzer. (Reproduced from Ref. 25, with kind permission.)

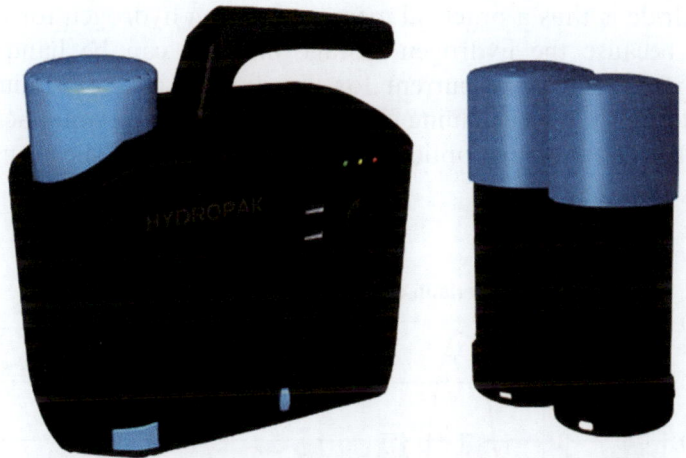

Figure 4.23   At 50–120 W, the HYDROPAK is the world's lowest cost fuel cell "off-grid" power device that combines PEM fuel cell technology with energy storage cartridges using sodium borohydride. (Reproduced from Ref. 25, with kind permission.)

The *HydroPak* (Figure 4.23) is the result of a collaborative effort with Millennium Cell,[29] a pioneering US fuel cell company that went out of business in 2008, to bring forth a low cost ($400) portable generator that uses $20 disposable cartridges, making it affordable and effective.

Millennium Cell developed a proprietary catalyst to generate hydrogen from a sodium borohydride dispersion in water in a controllable, heat-releasing reaction (Equation 4.1).

$$\underbrace{NaBH_4 + 2H_2O}_{\substack{\text{An energy-dense} \\ \text{water-based fuel} \\ \text{(i.e., 30 wt\% NaBH}_4 \\ \text{holds 6.7 wt\% H}_2\text{)}}} \rightarrow 4H_2 + \underbrace{NaBO_2(aq)}_{\substack{\text{Borate can be} \\ \text{recycled into} \\ \text{NaBH}_4}} + \underbrace{\sim 300\,kJ}_{\substack{\text{Exothermic} \\ \text{reaction} \\ \text{requires no heat} \\ \text{input}}} \tag{4.1}$$

Hydrogen is stored in a solid form at low pressure using metal hydride alloys that absorb H_2 molecules into their crystalline structure, and release them safely at low pressures. The fuel originating from Equation (4.1) is, at room-temperature, a non-flammable liquid under no pressure. No side reactions or volatile by-products are formed, whereas the H_2 gas generated has high purity (no CO or S) and is humidified (the heat generates some water vapor).

In general, storing hydrogen in metal hydrides creates the highest volumetric energy density of any form of hydrogen storage, even higher than liquid hydrogen (Figure 4.24).[30] Solid state storage of H_2 as sodium borohydride is thus a practical means of storing hydrogen for portable devices because the hydrogen storage material can be handled and transported easily. The current low production cost of sodium borohydride ($5\,\text{€}\,kg^{-1}$) is declining further, making this an economically and technically valid storage option for this and related applications of the H_2 fuel cell.

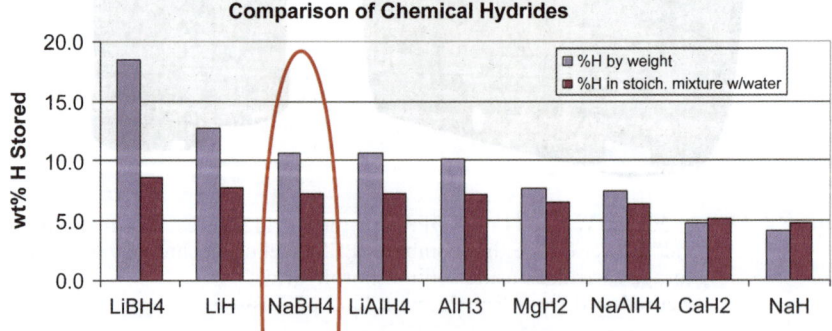

Figure 4.24 Hydrogen storage - Comparison of chemical hydrides in terms of weight percent.
(Reproduced from Y. Wu. *Hydrogen Storage via Sodium Borohydride. Current Status, Barriers, and R&D Roadmap*, GCEP – Stanford University, April 14–15, 2003, with kind permission.)

As a result, by adding water only, and plugging the *Hydrofill* into an electrical wall-socket (or a solar panel), consumers can automatically generate hydrogen and store it in a safe, solid form in *Hydrostik* cartridges. For example, 25 W DC power is enough to produce 10 L of hydrogen per hour, and to fill one *Hydrostik* cartridge with hydrogen stored in solid form.

With the *Hydrofill* desktop refueler (Figure 4.25), users no longer need to wait for the infrastructure for hydrogen to be developed, because water is ubiquitous and the device can use power either from the grid or from solar PV modules. Once full, the battery-like refillable *Hydrostik* can be unplugged from the *Hydrofill* and placed into a new portable power device named *Minipak* (or other fuel cell devices) to deliver 1.5 W via the USB port to power smartphones, lights and other devices.

Each *Hydrostik* can store 11 Wh of energy, which is enough for one or two charges of a 3G smartphone or two to three charges for the average cellphone. In comparison with current primary and rechargeable batteries, with 1 W power consumption, most AA batteries can only last 1 hour, while a *Hydrostik* has enough energy to last 10 hours. In other words:

1 Hydrostik = 10 disposable AA batteries at 1 W continuous power consumption.

Figure 4.25 The Hydrostik is competitive on cost/performance with existing battery devices, providing a low-cost (the whole system costs about $200) portable energy option for users.
(Reproduced from Ref. 25, with kind permission.)

Figure 4.26   The MINIPAK includes a passive air-breathing PEM fuel cell, able to draw hydrogen stored as solid hydride inside the specially designed HYDROSTIK cartridge.
(Reproduced from Ref. 25, with kind permission.)

Given that one cartridge can be refilled 100 times, the same *Hydrostik* over its lifetime replaces 1000 (10×100) disposable AA batteries.

The cartridges are inserted into the portable power charger and power extender *Minipak* (Figure 4.26), which is compatible with a variety of portable electronic devices and thus positioned to address the needs of users of power-hungry devices.

At the Consumer Electronics Show in 2012 in Las Vegas, the Swedish company myFC introduced its *PowerTrekk* (Figure 4.27) fuel cell charger (and *PowerPukk* fuelling cartridges), also aimed at the portable electronics market.[31]

Again, users simply insert a fuel "puck" and add water to allow rapid recharging of their cellphones, cameras and global positioning system (GPS) devices. Using sodium silicide (NaSi) storage, the fuel cartridges are activated by a small quantity of water and produce up to 4 L of hydrogen gas. Sodium silicide is a pyrophoric material. The fuel employed in the *PowerTrekk* is made of a specially formulated NaSi powder that is easily handled, does not react with dry oxygen, and absorbs moisture from air slowly without ignition.

Sodium silicide is a pyrophoric material that requires special storage and handling. The fuel developed by SiGNa Chemistry, Inc.[32] is made of Na_4Si_4 powders that react rapidly, but controllably, with water to produce hydrogen. The company has achieved the highest hydrogen

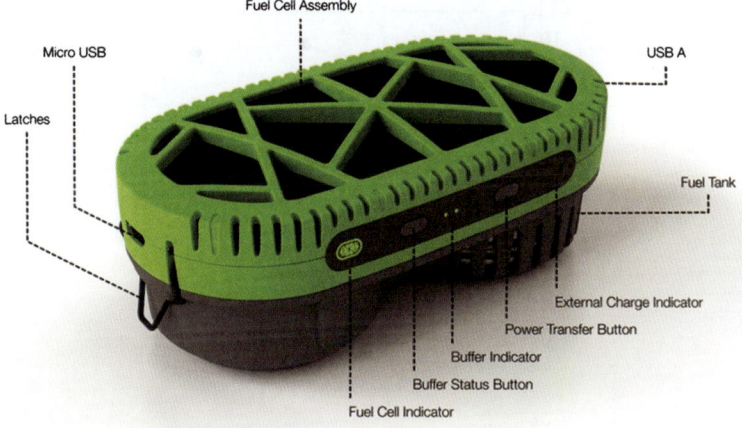

Fuel Cell Assembly

Micro USB

USB A

Latches

Fuel Tank

External Charge Indicator

Power Transfer Button

Buffer Indicator

Buffer Status Button

Fuel Cell Indicator

Figure 4.27 The new myFC portable, water-powered fuel-cell charger *PowerTrekk* can be used to recharge a smartphone. The PowerTrekk will be sold initially for $299, and will be on sale in the USA from May 2012. (Reproduced from Ref. 31, with kind permission.)

production yield of 9.5 wt%, *i.e.* 0.095 kg H_2 per 1 kg fuel, according to the reaction in Equation (4.2):

$$2NaSi(s) + 5H_2O(l) \rightarrow Na_2Si2O_5(aq.) + 5H_2 + Heat(\sim 175 \, kJ \, mol^{-1}) \quad (4.2)$$

In other words, the reaction produces five moles of hydrogen, almost instantaneously, from the reduction of five moles of water and only two moles of NaSi. As a result, sodium silicide provides greater power density per unit of weight than amorphous, sol-gel silica doped with sodium (Na-SG, Figure 4.28).[33]

At the heart of the fuel cell chargers is the *FuelCellStickers* proprietary technology. Made from foils and adhesives, these "stickers" – which are reliably and efficiently produced via a roll to roll, high volume manufacturing process – are assembled into a *flexible* assembly less than 2.75 mm thick. The company claims to be the world leader in micro fuel cell, planar technology featuring power density over 600 mW cm^{-3}.

The modular and scalable *FuelCellStickers* can also be assembled into "Blades" (Figure 4.29), providing higher power and efficient performance in a form factor less than 3 mm thick. These flexible components can be easily integrated on to the front or back covers, lids, bottoms and sides of mobile devices.

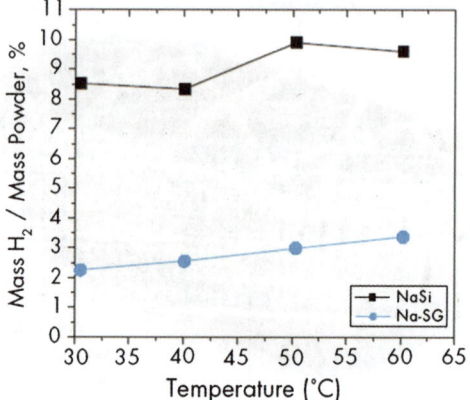

Figure 4.28   Production of hydrogen by the two SiGNa Chemistry materials. (Reproduced from Ref. 31, with kind permission.)

Figure 4.29   Flexible "Blades", made by myFC, are fuel cell components with a form factor less than 3 mm thick, ideally suited for customizable integration. (Reproduced from Ref. 31, with kind permission.)

4.7   An Insight into the Solar Hydrogen Economy

Although the English term "hydrogen economy", to describe a system of delivering energy using hydrogen, was coined by John Bockris, a former professor of chemistry at Texas A&M University, during a talk he gave in 1970 at General Motors, scientific interest in the solar

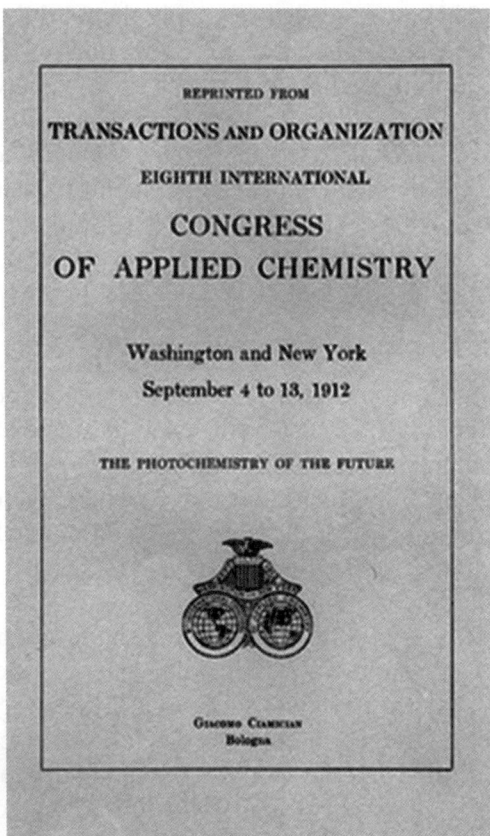

Figure 4.30 In a famous 1912 paper entitled "The Photochemistry of the Future", chemist Giacomo Ciamician foresaw the future energy transition from fossil to solar fuels.

hydrogen economy goes back to 1912, to the famous paper of Giacomo Ciamician entitled "The Photochemistry of the Future"[34] (Figure 4.30). This paper emphasized the need for an energy transition from fossil to solar fuels and foresaw that "our black and nervous civilization, based on coal", would be followed by "a quieter civilization based on the utilization of solar energy".

One hundred years later this prediction is eventually coming true. Indeed, a solar hydrogen economy finally makes sense, even if there are efficiency losses in storing (or liquefying) and delivering H_2,[35] because the available solar power is virtually unlimited *and*, concomitantly, the solar CSP and PV energy technologies, and the fuel cell and electrolyzer technologies, are eventually becoming low-cost and ubiquitous. Given that the sun delivers 5000 times our present global power needs, an area as small as $500 \times 500 \, km^2$ is needed to supply the world's energy needs (a tiny fraction of the world's desert area). Using mirrors, focused sunlight can heat water viably to generate electricity via a conventional steam turbine.

In a recent investigation based on order of magnitude calculations, without referring to environmental arguments but focusing on economic convenience only, Abbott has suggested that sunlight is the scalable source of power on which our future energy needs must rely, using low-tech CSP, where solar thermal collectors are preferred to PV solar cells.[36]

> The point about solar energy is that there is so much of it that you only have to tap 5% of it at an efficiency as tiny as 1% and you already have energy over 5 times the whole world's present consumption... There is so much solar that all you have to do is invest in the non-recurring cost of more dishes to drive a solar-hydrogen economy at whatever efficiency it happens to sit at.

Solar H_2 obtained from water *is* the fuel of the future because it solves the intermittency of supply of free solar energy, meeting one key requirement of modern societies: incessant flows of energy. Human activity and energy usage of course correlate significantly with the delivery of radiation from the sun, and solar hydrogen produced by water electrolysis is an excellent load-following clean technology.

For Abbott, combustion of solar H_2 should be preferred to hydrogen fuel cells because the latter are not scalable owing to the use of expensive membrane technology as well as expensive metal catalysts (platinum). By the same logic, CSP systems integrated with nanocatalysts are capable of splitting water *directly* and will have an immense impact on energy economics, because they require no electrolysis to provide affordable, renewable solar hydrogen with virtually zero CO_2 emissions. Such plants can offer new opportunities to regions of the world that have a huge solar potential, which can become important local producers of clean hydrogen.

We argue that the intrinsic versatility and the apparently endless falling price trend of PV electricity supports the use of solar PV stations to produce hydrogen locally from water electrolysis as the price of PV modules approaches a historic low of $0.5 \, W^{-1}$. If, for example, the roof of a $250 \, m^2$ (*ca.* 20 kWp) solar station in Austria can produce 823 kg of pure H_2 per year, in regions such as Sicily, where PV electricity has already reached grid-parity,[37] this figure should be almost doubled (70% more), further lowering the cost of solar hydrogen.

Figure 4.31 This home in New Jersey runs on solar power and stored solar hydrogen using only 56 solar PV panels on the garage roof and a small electrolyzer. In 2007, when the system was installed, the cost of a PV module was $7 W^{-1}$. Today, the same module sells at about $0.80 W^{-1}$. (Reproduced from Ref. 38, with kind permission.)

Smil agrees that transition to new energy sources is unavoidable, but remarks that even if a non-fossil world may be highly desirable, many decades will be needed for solar energy to capture substantial market shares on a global scale, because of the enormity of the requisite technical and infrastructure developments.[39]

Yet, low cost solar hydrogen will actually enable the shift to a distributed power distribution infrastructure, both in the developing countries of Africa and Asia, where hundreds of millions of people will start to use electricity *and* in affluent countries (Figure 4.31), where the price of fossil electricity and heat has been endlessly increasing for more than a decade.

Furthermore, new generation fuel cells and electrolyzers will be nanochemistry-based, requiring ever lower amounts of platinum to afford unprecedented performance in terms of power delivered and power density, and using much cheaper and readily available metal nanocatalysts such as nickel or molybdenum. It is perhaps not surprising

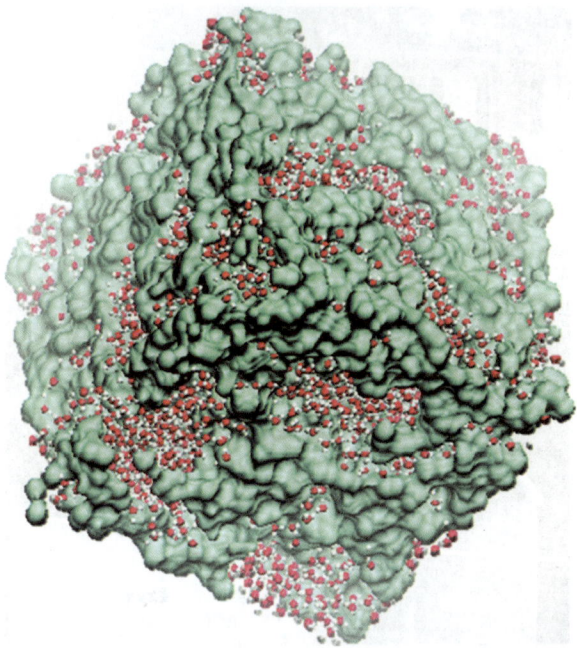

Figure 4.32   The tight-binding structure of APEs with high ion-exchange capacity for Ni-based fuel cells. In the self-cross-linking design used, a short-range cross-linker, tertiary amino groups, is grafted onto the quaternary ammonia polysulfone (QAPS), so that the cross-linking process can only occur during membrane casting. The self-cross-linked polysulfone is highly resistant to swelling: even at 80 °C, the membrane swells by less than 3%. (Reproduced from Ref. 38, with kind permission.)

that new platinum-free polymer electrolyte fuel cells (PEFC), comprised of alkaline polymer electrolytes (APEs, Figure 4.32), have been invented recently in China.[40]

In these innovative fuel cells the APE is as conductive and stable as its acidic Nafion membrane (a trademark of Du Pont) counterpart, which has been studied and used for decades in acidic fuel cells that use platinum catalysts. In detail, the new APEs are highly resistant to swelling and show excellent ionic conductivity at 80 °C, *i.e.* at typical temperatures for fuel-cell operation, which opens the route to low cost Ni-based fuel cells for a wide variety of commercial applications.[41]

Fuel cells, after all, are *already* selling without subsidies in many markets, and more than 700 000 fuel cells were sold cumulatively between 2005 and 2011 (Figure 4.33).[42] Portable, stationary and transport applications are all exhibiting growth, with the different electrolytes establishing their roles in the various sectors.[41] When low cost, Ni-based

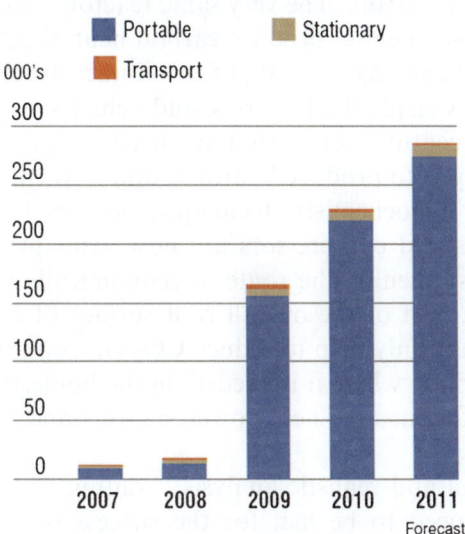

Figure 4.33 (heading)

Figure 4.33 Fuel cell shipments by application, 2007–2011.
(Reproduced from Ref. 38, with kind permission.)

catalysts are made available for alkaline fuel cells in the 10–200 kW power range, namely that of scooters, cars, trucks and boats, expect the transport column in Figure 4.33 to grow exponentially.

However, when are our global energy needs likely to be met using sunlight as energy input and water as raw material? Very shortly, we argue, within the next decade. This time, our argument is based on *economic* and *technological* reasoning. Where is, for example, the economic interest of countries such as China and the USA, the world's largest economies, to go ahead with dependence on foreign oil and natural gas when both countries own huge desert regions that are exceptionally suited to massive adoption of solar and solar hydrogen energy? Many countries, furthermore, have an urgent need to reduce the large public debt accumulated following public bailout of the financial system, create new jobs and reduce import of foreign oil and natural gas, which, even in a relatively small country like Italy, costs 63 billion Euros per year.[43]

Thermochemical water splitting using free and unlimited solar energy as the only energy input, via the highly scalable CSP technology coupled to nanocatalysis (see Chapter 3), will be used to produce *massive* amounts of carbon-free hydrogen by deploying very inexpensive solar reactors and mirrors on acres of vacant, non-productive land. To increase the scale of this system, it will be sufficient to add more solar reactors and more mirrors. The very same reactors will also be available to split carbon dioxide and produce carbon neutral solar fuels that can be piped into the existing natural gas or oil infrastructure for everyday use in homes, power plants, factories, and vehicles.

For almost a century, scientists have tried to split water cost effectively by electrolysis to produce hydrogen and oxygen. Today, however, using innovative nanochemistry techniques described in Chapter 2, low cost electrolyzers and compressors are now available, along with low cost PV modules, opening the route to economically viable distributed generation using part of the overall roof surface of existing buildings. Their products not only help to reduce CO_2 emission but will also help to produce electricity where it is needed, in the home. This also helps to avoid the need for massive new power stations and new transmission lines.

In a thoughtful and realistic analysis,[44] dating back to 2007, of the conditions that need to be met for the success of a hydrogen-based economy (Figure 4.34), Marbán and Valdés-Solís concluded that, first,

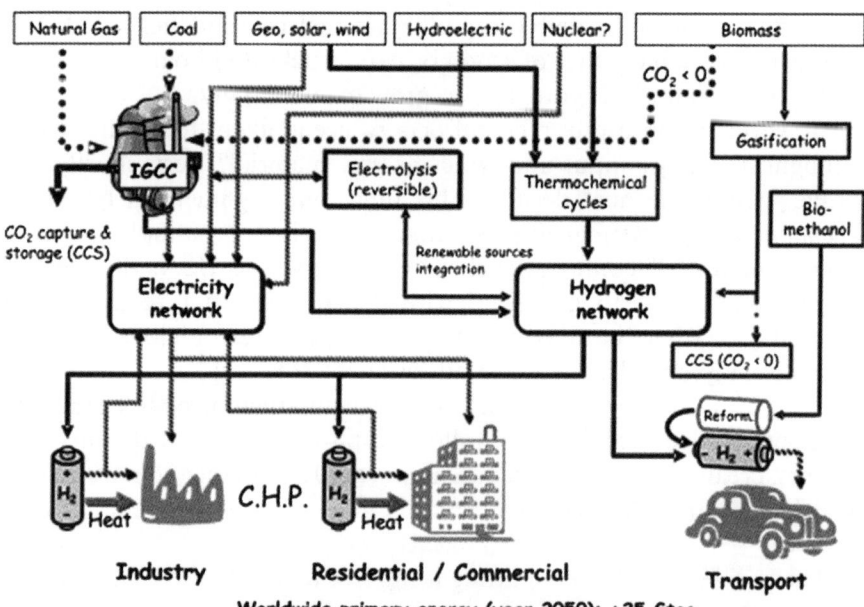

Figure 4.34 Expectations for the hydrogen society in the distant future. Renewable energies are intensified and hydrogen fuel cells are employed to achieve higher efficiencies.

international organizations should be strong enough to enforce the international agreements on global reductions in CO_2 emissions via a global emission market (an estimated cost of \$50 per ton of CO_2 was estimated to be sufficient to force energy companies to adopt carbon-free energy sources). Second, technological development should bring about a reduction in the costs of H_2 production, distribution, storage and utilization.

The latter change is actually taking place, as we assist in the continuous reduction in the cost of solar energy and related solar hydrogen technologies that goes along with the consistently high price of oil, which remains at $>$US\$90 per barrel, despite the prolonged global economic recession that started in 2008.[45]

Fuel cell prices, for example, are rapidly declining thanks to improving technology and scaling up of production. The year 2011 saw the first profitable fuel cell firms,[46] such as Electro Power Systems and Horizon Fuel Cells Technologies, mentioned above. The management of both these companies understood that in terms of the fuel-cell binomial it was the "fuel" and its availability on which they should focus their efforts. Their fuel cell devices integrate a water electrolyzer and a safe hydrogen storage system so that the user, by simply tapping water, becomes the "hydrogen infrastructure".[47]

As a result of these (and forthcoming) advances, we believe that fuel cells running on hydrogen will soon be found in many households, where solar H_2 will be produced locally using solar electricity.

As the number and reach of similar successful projects implemented worldwide grows, the huge potential of solar energy, in both developing countries endowed with ample solar energy, such as China, India and Brazil, and wealthy countries such as the USA and Australia, will become self-evident. We add to this our idea that further progress will increasingly make use of alternative forms of finance whose focus is on funding ethically and environmentally sound projects, and this will naturally have a great impact on the solar energy business. This will accelerate further widespread adoption of solar fuel energy that eventually will become the most economically and technically convenient fuel option, replacing our dependence on fossil fuels.

References

1. "Fuel Cell Drive Technology", www.daimler.com/technology-and-innovation/drive-technologies/fuel-cell (last accessed on 19/01/2012).
2. N. Armaroli and V. Balzani, *Ener. Environ Sci.*, 2011, **4**, 3193.

3. C.-J. Winter, *Int. J. Hydrogen Energy*, 2009, **34**, S1.
4. *EPA Mileage Estimates.* Honda FCX Clarity – Vehicle Specifications. American Honda Motor Company, 2010, http://automobiles. honda.com/fcx-clarity/specifications.aspx?group=epa (last accessed on 19/01/2012).
5. The DOE's Fuel Cell Power Model can be used for a preliminary evaluation of using fuel cells at the site location, www.hydrogen. energy.gov/fc_power_analysis.html (last accessed on 19/01/2012).
6. C. E. Thomas, Comparison of transportation options in a carbon-constrained world: hydrogen, plug-in hybrids, biofuels, *National Hydrogen Association Annual Meeting*, Sacramento (CA), 31 March 2008.
7. K. Ravindra, L. Bencs and R. Van Grieken, *Sci. Total Environ.*, 2004, **318**, 1.
8. S. Saulny, Thieves leave cars, but take catalytic converters, *The New York Times*, 29 March 2008, www.nytimes.com/2008/03/29/us/ 29converters.html (last accessed on 19/01/2012).
9. R. Freymann and H. Eichlseder, *The state of the art and future perspectives of the application of hydrogen I.C. engines*, BMW Group Clean Energy, 2006, http://althytude.info/fileadmin/user_upload/ documents/BMW_Freymann.pdf (last accessed on 19/01/2012).
10. S. Tamhankar (Linde), *Integrated Hydrogen Production, Purification and Compression System*, DOE Hydrogen Program, FY 2008 Annual Progress Report, www.hydrogen.energy.gov/pdfs/progress08/ ii_a_7_tamhankar.pdf (last accessed on 19/01/2012).
11. Hydrogen tries again, *The Economist*, 23 April 2010, www.economist. com/blogs/freeexchange/2010/04/hydrogen_and_future (last accessed on 19/01/2012).
12. BMW puts hydrogen test fleet on ice, *Deutsche Welle*, 7 July 2009, www.dw-world.de/dw/article/0,,4976630,00.html (last accessed on 19/01/2012).
13. *Honda Begins Operation of New Solar Hydrogen Station*, http:// world.honda.com/news/2010/c100127New-Solar-Hydrogen-Station/ (last accessed on 19/01/2012).
14. A. Ohnsman, Toyota advances hydrogen fuel cell plans amid industry's battery-car push, *Bloomberg*, 13 January 2011. www. bloomberg.com/news/2011-01-13/toyota-advances-hydrogen-plans-amid-industry-s-battery-car-push.html (last accessed on 19/01/2012).
15. The website includes extensive technical information and a detailed guide to standards and regulations.
16. A. Marisvensson, A. M. Svensson, S. Møller-Holst, R. Glöckner and O. Maurstad, *Energy*, 2006, **32**, 437.

17. According to the Pacific Northwest National Laboratory, for example, the idle off-peak grid capacity in the USA would be sufficient to power 84% of all US vehicles if they all were immediately replaced with electric vehicles: *Mileage from megawatts*, 2006, www.pnl.gov/news/release.aspx?id = 204 (last accessed on 19/01/2012).

18. C. E. Thomas, *Int. J. Hydrogen Energy*, 2009, **34**, 6005.

19. M. Weber and J. Perrin, in *Hydrogen Technology*, ed. A. Leon, Springer, Berlin, 2008, p. 129.

20. K. Gross Britta, I. J. Sutherland and H. Mooiweer, *Hydrogen fueling infrastructure assessment*, General Motors and Shell Hydrogen, 2007. www.h2andyou.org/pdf/GM-SH%20HYDROGEN%20INFRA%20PAPER.pdf (last accessed on 19/01/2012).

21. *Erstes Wasserstoff-Hybridkraftwerk in Brandenburg eröffnet*, 25 October 2011, www.verivox.de/nachrichten/erstes-wasserstoff-hybridkraftwerk-in-brandenburg-eroeffnet-80207.aspx (last accessed on 19/01/2012).

22. *Underground hydrogen storage*, http://en.wikipedia.org/wiki/Underground_hydrogen_storage (last accessed on 19/01/2012).

23. www.bloomenergy.com (last accessed on 19/01/2012).

24. T. Woody, A maker of fuel cells blooms in California, *Bits – New York Times*, 24 February 2010, http://bits.blogs.nytimes.com/2010/02/24/a-maker-of-fuel-cells-blooms-in-california (last accessed on 19/01/2012).

25. http://en.wikipedia.org/wiki/Bloom_Energy_Server (last accessed on 19/01/2012).

26. www.electrops.it (last accessed on 19/01/2012).

27. http://www.boconline.co.uk/products/products_by_type/industrial_products/hymera_generator.asp (last accessed on 19/01/2012).

28. www.horizonfuelcell.com (last accessed on 19/01/2012).

29. *Funeral for Millennium Cell*, 16 May 2008 www.hydrogencarsnow.com/blog2/index.php/fuel-cells/funeral-for-millennium-cell/ (last accessed on 19/01/2012).

30. B. Sakintuna, F. Lamari-Darkrim and M. Hirscher, *Int. J. Hydrogen Energy*, 2007, **32**, 1121.

31. www.myfuelcell.se (last accessed on 07/02/2012).

32. www.signachem.com (last accessed on 07/02/2012).

33. J. L. Dye, K. D. Cram, S. A. Urbin, M. Y. Redko, J. E. Jackson and M. Lefenfeld, *J. Am. Chem. Soc.*, 2005, **127**, 9338.

34. G. Ciamician, *Science*, 1912, **36**, 385.

35. U. Bossel, *Proc. IEEE*, 2006, **84**, 1826.

36. D. Abbott, *Proc. IEEE*, 2010, **98**, 1931.

37. P. Kanter, Industry group says solar to become cost-competitive in Italy next year, *Green blog, The New York Times*, http://greeninc. blogs.nytimes.com/2009/06/22/industry-group-says-solar-to-become-cost-competitive-in-italy-next-year/ (last accessed on 19/01/2012).
38. D. Biello, inside the solar-hydrogen house: no more power bills – ever, *Scientific American*, 19 June 2008, www.scientificamerican. com/article.cfm?id=hydrogen-house (last accessed on 19/01/2012).
39. (a) V. Smil, 21st century energy: Some sobering thoughts, *OECD Observer*, 2006, **258/59**, 22; www.oecdobserver.org/news/fullstory. php/aid/2083 (last accessed on 19/01/2012); (b) V. Smil, *Energy at the Crossroads*, MIT Press, Cambridge, MA, 2003.
40. S. Lu, J. Pan, A. Huang, L. Zhuang and J. Lu, *Proc. Natl. Acad. Sci. USA*, 2008, **105**, 20611.
41. J. Pan, C. Chen, L. Zhuang and J. Lu, *Acc. Chem. Res.*, 2012, **45**, 473.
42. *The Fuel Cell Today Industry Review 2011*, www.fuelcelltoday.com/ media/1351623/industry_review_2011.pdf (last accessed on 19/01/ 2012).
43. Reuters, *Italy's energy bill to rise steeply*, 16 June 2011. www.forexyard. com/en/news/Italys-energy-bill-to-rise-steeply-industry-2011-06-15T122932Z-UPDATE-1 (last accessed on 19/01/2012).
44. G. Marbán and T. Valdés-Solís, *Int. J. Hydrogen Energy*, 2007, **32**, 1625.
45. K. MacNamara, Manufacturing output suffers steepest fall in nearly 30 years, *The Independent*, 9 January 2009.
46. E. Wesoff, *Year-End Reflections on the Fuel Cell Industry in 2010*, 5 January 2011, www.greentechmedia.com/articles/read/Year-End-Reflections-on-Fuel-Cells-2010 (last accessed on 19/01/2012).
47. S. Pogutz, A. Russo and P. Migliavacca, *Innovation, Markets and Sustainable Energy – The Challenge of Hydrogen and Fuel Cell*, Edward Elgar Publishing, Cheltenham (UK), 2009.

Subject Index

Note: Italic page references indicate that the topic is mentioned only in a Figure or Table